Michael E. Auer · Kalyan Ram B. ·
Thrasyvoulos Tsiatsos
Editors

Artificial Intelligence in Education

Proceedings of the 16th IMCL Conference, Volume 2

Springer

Editors
Michael E. Auer
CTI Global
Frankfurt/Main, Germany

Kalyan Ram B.
Electrono Solutions Pvt. Ltd.
Bangalore, Karnataka, India

Thrasyvoulos Tsiatsos
Department of Informatics
Aristotle University of Thessaloniki
Thessaloniki, Greece

ISSN 2367-3370 ISSN 2367-3389 (electronic)
Lecture Notes in Networks and Systems
ISBN 978-3-032-24997-5 ISBN 978-3-032-24998-2 (eBook)
https://doi.org/10.1007/978-3-032-24998-2

This Springer imprint is published by the registered company Springer Nature Switzerland AG
The registered company address is: Gewerbestrasse 11, 6330 Cham, Switzerland

Lecture Notes in Networks and Systems 1954

The series "Lecture Notes in Networks and Systems" publishes the latest developments in Networks and Systems—quickly, informally and with high quality. Original research reported in proceedings and post-proceedings represents the core of LNNS.

Volumes published in LNNS embrace all aspects and subfields of, as well as new challenges in, Networks and Systems.

The series contains proceedings and edited volumes in systems and networks, spanning the areas of Cyber-Physical Systems, Autonomous Systems, Sensor Networks, Control Systems, Energy Systems, Automotive Systems, Biological Systems, Vehicular Networking and Connected Vehicles, Aerospace Systems, Automation, Manufacturing, Smart Grids, Nonlinear Systems, Power Systems, Robotics, Social Systems, Economic Systems and other. Of particular value to both the contributors and the readership are the short publication timeframe and the world-wide distribution and exposure which enable both a wide and rapid dissemination of research output.

The series covers the theory, applications, and perspectives on the state of the art and future developments relevant to systems and networks, decision making, control, complex processes and related areas, as embedded in the fields of interdisciplinary and applied sciences, engineering, computer science, physics, economics, social, and life sciences, as well as the paradigms and methodologies behind them.

Indexed by SCOPUS, EI Compendex, INSPEC, WTI Frankfurt eG, zbMATH, SCImago.

All books published in the series are submitted for consideration in Web of Science.

For proposals from Asia please contact Aninda Bose (aninda.bose@springer.com).

Preface

IMCL2025 was the 16th edition of International Conference on Intelligent Multimodal Communication and Learning Technologies.

IMCL2025 is part of an international initiative to promote technology-enhanced learning and online engineering worldwide. Intelligent multimodal communication and learning technologies integrate multiple input and output modalities—such as speech, text, gestures, vision, and haptics—enhanced by artificial intelligence (AI) to create more effective, adaptive, and interactive learning environments. These technologies aim to improve human-computer interaction (HCI), enhance accessibility, and support personalized education.

IMCL2025 conference covered all aspects of multimodal communication as well as the emergence of learning technologies, infrastructures and services and their implications for education, business, governments and society. IMCL2025 promoted the development of e-learning, to provide a forum for education and knowledge transfer, to expose students to latest ICT technologies and encourage the study and implementation of multimodal communication applications and technologies in teaching and learning. The conference also stimulated critical debate on theories, approaches, principles, and applications of learning technologies among educators, developers, researchers, practitioners and policymakers.

IMCL2025 has been co-organized by International Association of Online Engineering, Atria University, India, and Aristotle University of Thessaloniki, Greece, from 19 to 21 November 2025.

This year's theme of the conference was "**Artificial Intelligence in Education**".

Again, outstanding scientists from around the world accepted the invitation for keynote speeches:

- **Krishna Vedula**, Founder and Executive Director, IUCEE and Dean Emeritus, University of Massachusetts Lowell, USA: *Strengthening Learner-Centric Pedagogy using Technological Tools.*
- **Sunder Raju**, Chancellor of Atria University, Chairman of Atria Power, Bengaluru, India: *The Age of Agents: AI's Encroachment and Humanity's Response.*
- **Reinhard Langmann**, Chairman of EduNet World Association, Germany: *Global Learning Networks and Smart Engineering Education: Empowering the Future through Remote Labs, AI and Collaboration.*
- **Gustavo R. Alves**, Portuguese National Agency for Science and Technology: *AI Impact on Traditional and Non-Traditional Instructional Laboratories.*

Furthermore, interesting workshops and tutorials have been organized:

- **Introducing Computing Tool in Engineering Education: An Example through Python for Signal Processing Concepts**, by Kanmani Buddhi, BMS College of Engineering.

- **Industrial Realities: Robotics, Multimodal Communication, & Edge AI for Future Workforces** by Vijai Simha, Atria University.
- **Global Remote Lab Initiative** by Reinhard Langman, Edunet World Association and Michael Auer, IAOE.

Since its beginning this conference has been devoted to new approaches in learning with a focus on mobile learning, mobile communication, mobile technologies and engineering education.

We are currently witnessing a significant transformation in the development of Intelligent Multimodal Communication and Learning Technologies.

Therefore, the following main topics have been discussed during the conference in detail:

A. **Multimodal Learning Technologies.**

- Multimodal Inputs and Outputs.
- Conversational AI.
- Context-Aware Interfaces.
- Gesture-Based Controls.
- Multilingual Capabilities.
- Augmented-, Virtual-, Mixed- and Cross-Reality.
- AI and Machine Learning for Adaptive Learning.
- Multimodal Content Delivery.

B. **Intelligent Learning Technologies.**

- Artificial Intelligence in Education.
- Learning analytics
- Human-Computer Interaction (HCI) and Smart Interfaces.
- Wearables and Internet of Things (IoT).

C. **Applications of Intelligent Learning Technologies.**

- Personalized Education.
- Corporate Training and Professional Development.
- Assistive Learning Technologies.
- Smart Classrooms and Virtual Learning Environments.
- Gamification and Serious Games.
- Online Learning and E-Learning Platforms.

D. **Challenges and Future Directions.**

- Data privacy and security concerns.
- Bias in AI algorithms impacting fairness in education.
- Digital Divide.

- 5G and Edge Computing.
- AI-driven emotional intelligence and metaverse learning spaces.

The following Special Sessions have been organized:

- **Building AI Resilience in Education: AI-Native Pedagogies and Curricular Processes (BAIRE)**, Chair: Prof. Varghese Panthalookaran, Rajagiri School of Engineering and Technology, India.
- **Artificial Intelligence in Learning and Assessment (AILA-25)**, Chairs: Jyotsna Kumar Mandal, University of Kalyani, Kalyani, India, and Rajeev Chatterjee, National Institute of Technical Teachers Training and Research, Kolkata, India.

As submission types have been accepted:

- Full Paper, Short Paper, Work in Progress (with in person or distant/pre-recorded presentation).
- Poster.
- Special Sessions.
- Round Table Discussions, Workshops, and Tutorials.

All contributions were subjected to a double-blind review. The review process was very competitive. A total of **157** submissions were reviewed. A team of **61** reviewers carried out this demanding task. Our special thanks go to all of them.

Due to the time and conference schedule restrictions – only the best **51** submissions could finally be accepted for presentation.

The following papers were distinguished as **best papers**:

- **Category "Full Paper"**: "Improving Ethical Reasoning with an LLM: A Scenario-Based Tool for Higher Education", by Stefania Zourlidou, Peter Oloke, Institute of Web Science and Technologies (WeST), Germany.
- **Category "Short Paper"**: "Growing Adoption of Conversational Agents for Workplace Learning: A Scoping Review" by Eirini Voutsa, and Stavros Demetriadis, Aristotle University of Thessaloniki, Greece.
- **Category "Work-in-Progress"**: "Work-in-Progress: Curating Primary Sources for Civic Themes: VR-Ready Digital Galleries in Teacher Education" by Esra Özcan, Tülin Haşlaman, and Gökçe Nur Yılmaz, Ted University, Turkey.

Our conference attracted more than **144** participants from **30** countries.

We thank **Sebastian Schreiter** for the technical editing of these proceedings.

IMCL2026 will be co-organized in Athens, Greece, by International Association of Online Engineering, UNIC Athens, and Aristotle University of Thessaloniki.

Michael E. Auer
IMCL Steering Committee Chair

Thrasyvoulos Tsiatsos
IMCL General Chair

Kalyan Ram B.
IMCL2025 Chair

Committees

Steering Committee Chair

Michael E. Auer	CTI, Frankfurt/Main, Germany

General Conference Chair

Thrasyvoulos Tsiatsos	Aristotle University of Thessaloniki, Greece

IMCL2025 Chair

Kalyan Ram B.	Indxo AI, Bangalore, India

International Chairs

Alaa Ashmawy	American University Dubai (Middle East)
Uriel Cukierman	University of Buenos Aires, Argentina (South America)
Samir A. El-Seoud	The British University in Egypt (Africa)
David Guralnick	Kaleidoscope Learning New York, USA (North America)
Alexander Kist	University of Southern Queensland, Australia (Australia/Oceania)
Tiia Rüütmann	Technical University Tallinn (Europe)

Technical Program Chairs

Thrasyvoulos Tsiatsos	Aristotle University of Thessaloniki, Greece
Sebastian Schreiter	IAOE, France

Workshop, Tutorial, and Special Session Chair

Thrasyvoulos Tsiatsos	Aristotle University of Thessaloniki, Greece

Publication Chair and Web Master

Sebastian Schreiter	IAOE, France

Program Committee Members

Abul Azad	Northern Illinois University, USA
Abdallah Al-Zoubi	PSUT Amman, Jordan
Agisilaos Konidaris	Ionian University, Greece
Anastasios Economides	University of Macedonia, Greece
Apostolos Gkamas	University Ecclesiastical Academy of Vella of Ioannina, Greece
Christos Katsanos	Aristotle University of Thessaloniki, Greece
Christos Douligeris	University of Piraeus, Greece
Daphne Economou	University of Westminster, UK
David Jennings	University College, Ireland
Demetrios Sampson	University of Pireaus, Greece
Despo Ktoridou	University of Nicosia, Cyprus
Dieter Wuttke	Technical University Ilmenau, Germany
Dominik May	University of Wuppertal, Germany
Doru Ursutiu	University Transylvania Brasov, Romania
George Magoulas	Birkbeck College, UK
George Palaigeorgiou	University of Western Macedonia, Greece
Giasemi Vavoula	University of Leicester, UK
Gustavo Alves	Polytechnic of Porto, Portugal
Juergen Koeberlein-Kerler	Wuerzburg, Germany
Kostas Apostolou	McMaster University, Canada
Krishna Vedula	IUCEE Hyderabad, India
Maiga Chang	Athabasca University, Canada
María Isabel Pozzo	National Technological University, Argentina
Manuel Castro	Universidad Nacional de Educación a Distancia, Spain
Margareth Gfrerer	University Dire Dawa, Ethiopia
Michail Giannakos	Norwegian University of Science and Technology, Norway
Monica Divitini	Norwegian University of Science and Technology, Norway
Nektarios Moumoutzis	Technical University of Crete, Greece
Panarit Sethakul	KMUTNB, Thailand
Petros Lameras	The Serious Games Institute, UK

Rajeev Chatterjee	National Institute of Technical Teachers' Training & Research, Kolkata, India
Santi Caballé	Open University of Catalonia, Spain
Sarmad Ahmed Shaikh	Sindh Madressatul Islam University, Karachi, Pakistan
Stelios Xinogalos	University of Macedonia, Greece
Stavros Nikou	University of Strathclyde, UK
Stamatios Papadakis	The University of Crete, Greece
Tharenos Bratitsis	University of Western Macedonia, Greece
Ting-Ting Wu	National Yunlin University of Science and Technology, Taiwan
Valérie Varney	TH Cologne, Germany

Contents

Personalized Education

Augmented-, Virtual-, Mixed- and Cross-Reality

Integrating Augmented Reality and Machine Learning for Suggestive Dermatological Diagnostics in Black African Populations

Olamiposi Olaiya(✉), Gerry Chan, and Rita Orji

Faculty of Computer Science, Dalhousie University, Halifax, NS, Canada
ol264778@dal.ca

Abstract. In this research, we present SkinVista, a novel mobile application integrating augmented reality (AR) and machine learning (ML). SkinVista enhances dermatological diagnostics for Black African populations by addressing diagnostic disparities in darker skin tones. This exploratory study evaluated SkinVista's efficacy through a pilot with six participants, focusing on usability, usefulness, technology acceptance, and recommendation likelihood. Using a fine-tuned MobileNetV2 model trained on 960 images per class of eight common skin conditions (e.g., acne, eczema), the app provided suggestive diagnoses with AR-guided image capture. Quantitative results showed that the app is usable, useful and informative. Results also showed that participants found the app easy to use and relevant for delivering AI-generated diagnostic suggestions with educational feedback, and they were willing to recommend it to others. These findings highlight SkinVista's potential to empower users with accessible skin health insights while demonstrating the need for improved user interface design to enhance adoption and satisfaction.

Keywords: Augmented Reality (AR) · Machine Learning (ML) · Dermatological Diagnostics · Black African Populations

1 Introduction

The skin is the body's largest organ, serving as a physical barrier between the internal organs and the external environment. It is important for protecting the body from harmful elements such as bacteria, fungi, viruses, and chemical insults. Structurally, the skin consists of two primary layers: (1) the epidermis, and (2) the dermis. The epidermis, the outermost layer, is composed of stratified squamous epithelial cells and varies in thickness from approximately 75-150μm depending on the body region [1, 2]. It protects the human body from dehydration and other physical and chemical changes in the environment [3]. The second major layer of skin is the dermis, which exists underneath the epidermis. The dermis is a connective tissue that is thicker than the epidermis and contains the major mass of human skin. The dermis contains extensive nerve and vascular networks, sweat glands, and keratinized structures, which include sebaceous and hair glands [4]. Together, these layers contribute to thermoregulation, immune defense, and sensory perception.

M. E. Auer et al. (Eds.): IMCL 2025, LNNS 1954, pp. 3–12, 2026.
https://doi.org/10.1007/978-3-032-24998-2_1

Skin color is determined by the amount of melanin produced by melanocytes in the epidermis. The Fitzpatrick scale (Types I–VI) classifies skin based on its response to ultraviolet radiation, ranging from fair to deeply pigmented. This classification is significant in dermatology, as variations in skin type influence the presentation and diagnosis of skin conditions. However, skin diseases are common and can range from mild irritations to life-threatening conditions. Without proper treatment, some conditions can lead to long-term health complications. Unfortunately, medical research has shown that dermatology often lacks diverse representation, which can make it more difficult to accurately diagnose and treat skin conditions in individuals with darker skin tones [2, 5]. According to the World Health Organization (WHO), skin diseases are widespread in many parts of the world. In African communities, between 35% and 87% of people are affected by these conditions [6]. Access to specialized dermatological care is limited for many, hindering appropriate treatment.

Augmented Reality (AR) and Machine Learning (ML) are increasingly recognized as useful technologies in the healthcare domain by helping in the diagnosis of various conditions, supporting personalized treatment planning, and enhancing patient education through interactive and immersive experiences [7]. AR allows users to see helpful information overlaid on real-world images, while ML helps computers recognize patterns and make predictions. Researchers have explored AR in surgeries and rehabilitation [8] and found that combining ML with AR improves accuracy in medical diagnoses [9]. However, existing Artificial Intelligence (AI) tools often make errors when analyzing darker skin tones, highlighting the need for more inclusive technology [5]. Although there have been efforts to improve dermatological diagnostics in Africa, many challenges remain unaddressed. For example, some researchers have developed web-based detection systems [10] and deep-learning diagnostic tools for specific skin conditions [11]. However, there has been little research on combining AR and ML to create an easy-to-use, culturally relevant mobile application designed specifically for Black Africans.

In this research, we contribute to ongoing efforts on the design and use of technology to educate communities about skin health and improve early detection of skin conditions [12–14]. We present a novel, AR- and ML-enhanced mobile application designed to improve awareness, support early diagnosis, and expand access to dermatological guidance for Black African populations. By focusing on common conditions in darker skin tones, the app aims to reduce diagnostic disparities and promote health literacy. Through AR-guided scanning, real-time feedback, and clear explanations, SkinVista integrates visual learning and personalized data to enhance skin health knowledge and support proactive self-care.

2 Methodology

2.1 Study Design and Participants

This exploratory study was designed to evaluate the effectiveness of SkinVista, a mobile application integrating AR and ML for suggestive dermatological diagnostics tailored to Black African populations. The pilot study involved six participants, randomly selected to ensure a diverse representation of individuals with darker skin tones affected by dermatological conditions, including acne, eczema, atopic dermatitis, keloids, fungal infections, pseudofolliculitis barbae, ringworm, and vitiligo. Participants had no prior dermatological training, simulating typical end-users of the application. The small cohort size was chosen to facilitate in-depth qualitative feedback and iterative refinement of the application in a controlled setting.

2.2 Prototype Design

First, we created a low-fidelity prototype using Figma (see Fig. 1) to design the layout and structure of the app's interface elements. Once the basic screens were in place, we developed a high-fidelity interactive prototype (see Fig. 2), also using Figma, to simulate the user experience. SkinVista guides users as they scan their skin by providing on-screen, real-time tips to help them take clear pictures of affected skin areas. After an image is captured, the app sends it to the ML model, which then returns a suggested diagnosis.

Fig. 1. Initial low fidelity prototype of SkinVista.

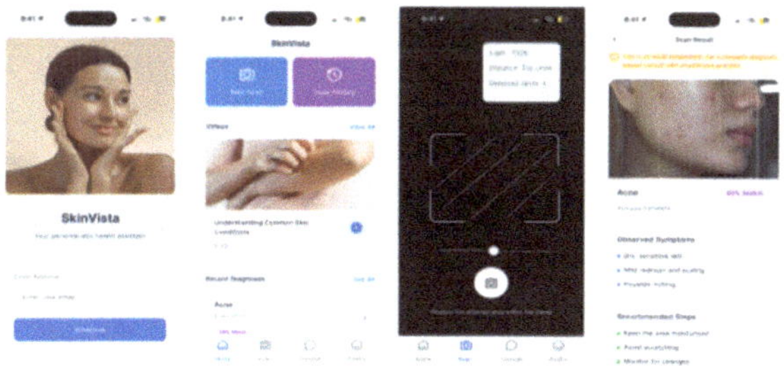

Fig. 2. Interactive high-fidelity prototype of SkinVista.

2.3 Tools and Techniques

The SkinVista application was developed using Flutter for the front-end mobile interface, ensuring cross-platform compatibility for iOS and Android devices. The backend was built with Laravel, a PHP framework, to handle image processing, ML model integration, and data management. The ML model was constructed using TensorFlow with a MobileNetV2 architecture, fine-tuned on a dataset of 960 images per class, covering eight common dermatological conditions prevalent in darker skin tones: acne, eczema, atopic dermatitis, keloids, fungal infections, pseudofolliculitis barbae, ringworm, and vitiligo.

The dataset was curated to focus on darker skin tones, addressing the gap in diverse representation. Data augmentation techniques, including random flips, brightness, contrast, hue, and saturation adjustments, were applied to enhance model robustness.

The AR component was implemented using Flutter's AR capabilities, leveraging the device's camera to overlay real-time visual prompts. These prompts guided users to adjust lighting, distance, and angle for optimal image capture, thereby improving image quality for ML analysis. The ML model processed images at a resolution of 224 x 224 pixels, outputting a suggested diagnosis along with a corresponding confidence score. A confidence threshold of 0.5 was set to filter low-confidence predictions, returning "no diagnosis" when applicable to reduce false positives. Before deployment, dermatologists reviewed and validated the ML model's predictions to ensure diagnostic accuracy aligned with expert assessments, after which the application operated independently, providing diagnoses directly to users based on submitted images. Figure 3 illustrates the visual pipeline of SkinVista's AR-guided image capture, ML model analysis, and diagnosis display.

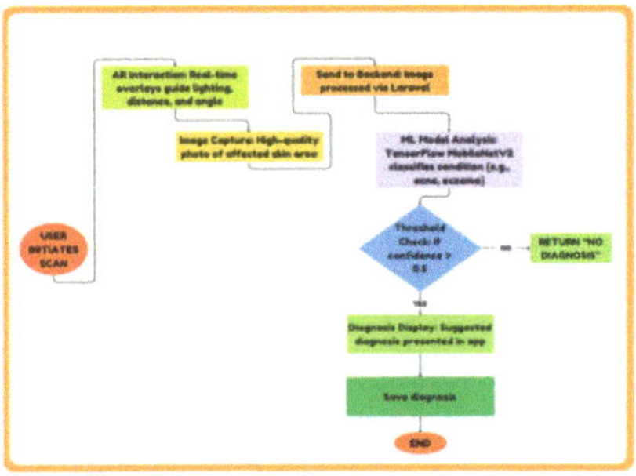

Fig. 3. SkinVista visual pipeline.

2.4 Procedures

The study was conducted over multiple sessions, and ethical approval was obtained from our institutional ethics review board before participant recruitment. Before any participant interaction, the ML model was validated by dermatology experts to ensure diagnostic consistency across the target skin conditions. A stratified sample of 10 images per condition was drawn from the training dataset and independently reviewed by two board-certified dermatologists. The dermatologists then (a) diagnosed each image based on visual assessment, (b) compared their diagnosis with the ML model's predictions, and (c) rated the model's accuracy using a 5-point Likert scale. This process ensured that the model's output was clinically plausible before being deployed for participant testing. Each session began with a five-minute tutorial on using SkinVista, delivered through an onboarding screen to ensure consistent user orientation. Participants used their smartphones to capture images of skin areas affected by one of the target conditions. The AR system provided real-time guidance to ensure high-quality image capture. Each participant submitted one image, which was processed by the ML model to generate a suggested diagnosis. The results were presented directly to the participant through the app, relying on the pre-validated ML model without real-time dermatologist involvement. Following the interaction, participants completed a structured questionnaire, designed to capture their experiences regarding usability, trust in the ML suggestions, and the effectiveness of AR guidance. A semi-structured interview was also conducted afterward to obtain qualitative feedback and deeper insights into participants' experiences.

A structured questionnaire was developed using a combination of established scales and custom survey items. The instrument covered four major criteria: usefulness and informativeness (e.g., relevance and helpfulness of the app), technological acceptance (e.g., ease of use, functional value), usability, and likelihood of recommendation (on a 10-point scale). To measure usefulness and informativeness, we used the Usefulness and Informativeness Scale [15], which evaluates the relevance and helpfulness of digital tools in supporting user tasks. Technology acceptance was assessed using the Technology Acceptance Scale [16], which captures constructs such as ease of use and perceived functional value. To measure usability, we used the System Usability Scale (SUS) [17], where participants indicated their level of agreement on a 5-point Likert scale ranging from 1 (strongly disagree) to 5 (strongly agree). The SUS is a widely used, validated questionnaire that provides a reliable measure of a system's overall usability, including mobile health applications [18]. Responses for other constructs were also collected using a 5-point Likert scale (1 = "strongly disagree" to 5 = "strongly agree"), and scores were aggregated for analysis.

2.5 Analysis

Quantitative analysis involved two components: (1) ML model performance, and (2) participant questionnaire responses. For the ML model, performance metrics, including precision, recall, F1-score, and Cohen's Kappa score, were computed using the test set of balanced images. A confusion matrix (Fig. 4) was also generated to identify patterns in misclassifications, with misclassified images logged for further analysis.

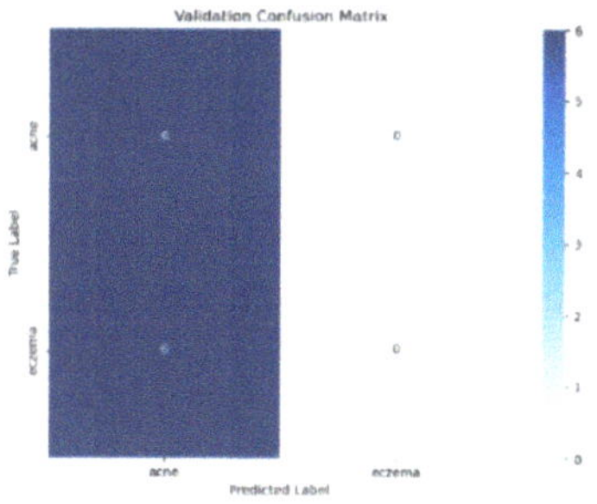

Fig. 4. Confusion matrix illustrating model performance in distinguishing between acne and eczema.

The confusion matrix shows that the model correctly identified all cases of acne but failed to identify any eczema cases, misclassifying them entirely as acne. This highlights the model's limited ability to differentiate between visually similar conditions, emphasizing the need for additional training data and refinement to improve classification accuracy. Participants were first given access to the SkinVista application and asked to interact with its features, including scanning for skin conditions and reviewing diagnostic feedback. Immediately after use, they completed the questionnaire based on their experiences. Data was recorded anonymously to ensure confidentiality and reduce response bias. Qualitative feedback from the semi-structured interview was analyzed using an inductive approach [19] to identify patterns in user experience, focusing on usability, trust in the ML suggestions, and the effectiveness of AR guidance, as they emerged from the data. Descriptive statistics were used to interpret trends and variations across user responses.

3 Results

3.1 Demographic Characteristics

A total of six respondents participated in the study. Most participants (five out of six) were between the ages of 22 and 35 years old, while one respondent was in the 15–21 age range. These age groups represent a demographic typically more concerned with skin appearance and health, aligning well with the target user base for the SkinVista application. Gender distribution was skewed, with five male participants and one female participant. While this sample provided initial insights. Quantitative analysis of user responses to the SkinVista app reveals a generally positive perception across core domains, with particularly strong scores in usefulness and informativeness, technology acceptance, and likelihood to recommend. However, slightly lower scores in system usability indicate areas for further improvement, particularly around user experience design and interface complexity.

3.2 Usefulness and Informativeness

Participants demonstrated a high appreciation for the app's core value. This indicates that most users found SkinVista helpful for diagnosing and understanding their skin conditions, as well as potentially enabling early detection. The strong performance in this category reinforces the app's relevance in a personal health management context and suggests that it effectively meets a key user need for reliable dermatological insights. Figure 5 shows the usefulness (5a) and the informativeness (5b) of the SkinVista app.

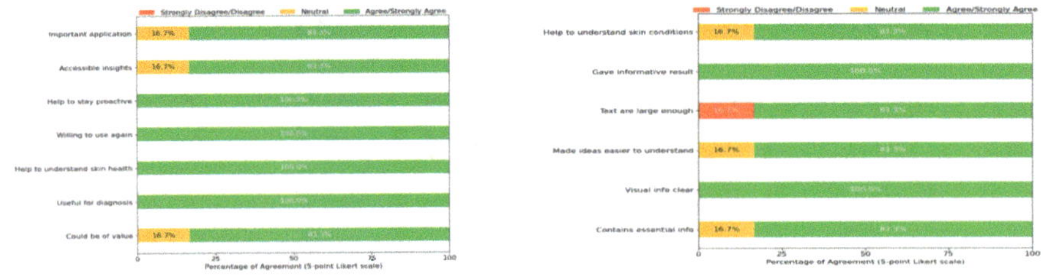

(a) Usefulness

(b) Informativeness

Fig. 5. Distribution of usefulness and informativeness ratings displayed as 100% stacked bar charts, based on evaluations using the IMI [15].

3.3 Technology Acceptance

Technology acceptance also scored favorably. Participants largely agreed that the app is beneficial, contextually relevant, and easy to use. These results suggest that SkinVista successfully aligns with users' expectations and reflects core principles of the Technology Acceptance Scale, wherein perceived usefulness and ease of use drive positive attitudes toward technology adoption. Figure 6 shows the technological acceptance of the SkinVista app.

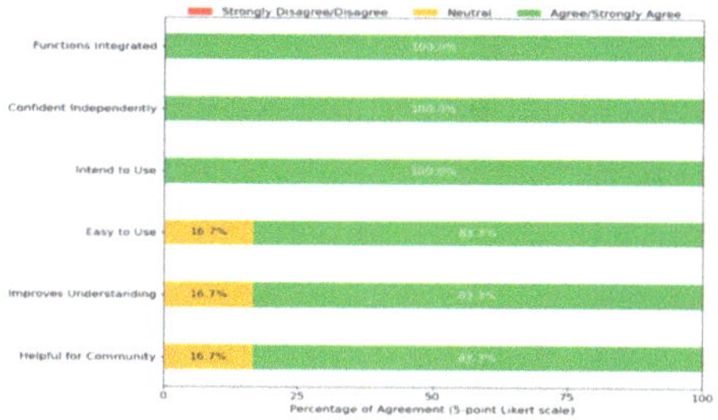

Fig. 6. Technological acceptance of the SkinVista app.

3.4 Recommendation Behavior

Despite some usability challenges, participants expressed a strong intent to recommend the SkinVista app to others. Most respondents rated their likelihood of recommending the app as 8 or higher on a 10-point (1 = "very unlikely" to 10 = "very likely") scale. This suggests that the SkinVista app is generally easy to use and well-received, with most users finding it simple, consistent, and confidence-inspiring without needing technical help. However, some users were less certain about using it frequently, which could be a challenge for those less comfortable with technology. The high recommendation scores show that users value the app's usefulness, which may encourage others to try it. To make the app even better, improving the interface to make it simpler and easier to learn could increase user satisfaction and help it reach more people through recommendations.

3.5 Comparative Analysis of System Usability and Recommendation Scores

As shown in Fig. 7, a comparative analysis of SUS and recommendation scores for the SkinVista app ($n = 6$) suggests a positive association between usability and user advocacy. All participants with SUS scores above the midpoint of 50 ($M = 82.5$, $SD = 6.4$) provided recommendation scores of 7 or higher ($M = 8.83$, $SD = 0.75$), with scatter plot clustering in the high-SUS, high-recommendation region indicating that users who found the app easy to use and intuitive were more likely to advocate for it. While these results suggest that users generally found the app usable and helpful, there may be usability challenges for individuals who are less technically savvy. This highlights the importance of refining the interface to ensure accessibility and ease of use for a broader user base.

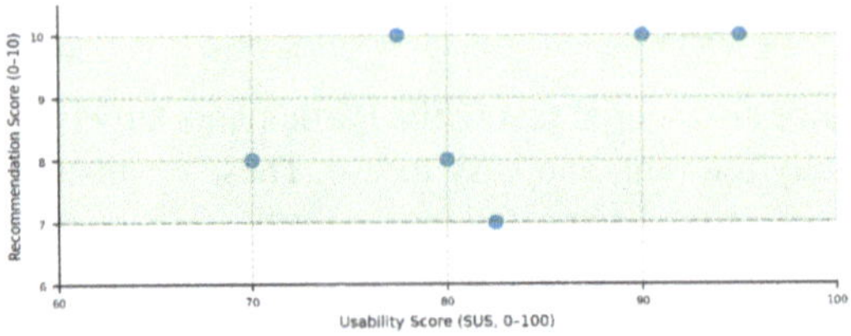

Fig. 7. A comparative analysis of SUS and recommendation scores.

3.6 Qualitative Results

To obtain additional feedback and a broader perspective of participants' experiences, we conducted short semi-structured interviews with the 6 participants who completed the study to understand the users' views about SkinVista. Interview transcripts were analyzed using Braun and Clarke's [20] six-phase thematic analysis approach. The analysis revealed 3 main themes related to participants' perception about the app: (1) accessibility and ease of use, (2) perceived usefulness for early detection and self-awareness, and (3) concerns and suggestions for improvement. The extracts are presented with minor spelling and grammatical corrections.

- **Theme 1: The app is easy to use and accessible.** Participants generally found the app straightforward and convenient to use. For example, one participant said, *"I don't usually like skin apps because they can be confusing, but this one was very clear. I just took the picture and got the result without stress"* [P3]. Another participant commented the *"The icons are very easy to understand. Even someone who is not very good with technology will understand the app's flow"* [P5]. These comments confirm that the design choices contributed to a positive user experience.
- **Theme 2: Useful for self-awareness.** Many participants noted that SkinVista increased their awareness of skin health and could be a helpful tool for early suggestive diagnosis. One participant said, *"I didn't know the name of the skin disease I had, your app made me know the name"* [P2]. Another participant expressed that *"It gave me an idea of whether I should go see a doctor or not"* [P4]. In general, these reflections highlight the app's potential to support proactive health-seeking behaviors.
- **Theme 3: Concerns and suggestions for improvement**. While the general response was positive, participants also expressed concerns and provided suggestions. For example, one participant stated, *"The result is good, but sometimes I might need more context for the recommended steps"* [P6]. Another participant said, *"The AR guidance features freeze after a long duration of using the camera"* [P1]. These responses emphasize the importance of providing more detailed feedback and improving technical performance to strengthen usability.

Overall, the qualitative findings indicate that SkinVista was perceived as accessible and useful, with participants recognizing its potential to increase the awareness of skin health and support timely medical consultation.

4 Conclusion, Limitations, and Directions for Future Work

In closing, this exploratory study investigated how SkinVista can support early detection and awareness of skin conditions through the integration of AR and ML. Designed specifically for Black African populations, the app offers an interactive and educational experience that empowers users to better understand and manage their skin health. By delivering personalized feedback, visual guidance, and culturally relevant content, SkinVista exemplifies how intelligent mobile technologies can advance equitable healthcare and foster digital health literacy in underserved communities. Beyond its technical contributions, SkinVista also demonstrates how ML and AR can transform health diagnostics into interactive learning experiences. Through real-time prompts and simple explanations, the app encourages users to engage more deeply with their health, making skin care more understandable and actionable. In doing so, it connects with broader efforts to make learning more accessible, personalized, and embedded in everyday life, especially for users with limited access to formal medical guidance.

Additionally, pilot studies often involve small sample sizes [21], which can limit the extent to which observed trends can be generalized. As such, there is the need for larger studies to validate the relationship between usability and behavioral intentions. Nevertheless, this pilot study serves as a valuable foundation for validating key features and laying the groundwork for future development. Future research will aim for at least 30 participants to achieve greater statistical significance and prioritize inclusion of a more gender-balanced and demographically diverse cohort to enhance representativeness. Moving forward, the following app versions will focus on enhancing navigation through the application, improving the user experience, and conducting user testing more broadly. We plan to enhance the onboarding screen to more clearly identify the navigation items and describe the experience in each section (e.g., Home, Scan, Results, Profile) so the user has greater awareness of the menu items available to them. To improve the image capture experience, we will replace text instructions with short animations or picture guides showing how to position the device and take a photo. Usability testing will expand to include older adults and individuals with low-tech experience, using think-aloud sessions to gather real-time feedback during app use. These changes aim to strengthen the app as an accessible, informative, and impactful skin health resource that promotes equitable digital health engagement.

References

1. Bonifant, H., Holloway, S.: A review of the effects of ageing on skin integrity and wound healing (2019)
2. Abbas, M., Imran, M., Majid, A., Ahmad, N.: Skin Diseases Diagnosis System Based on Machine Learning. J. Comp. Biomed. Info. 4 (2022)
3. Members of the South West Regional Wound Care Program's Clinical Practice and Knowledge Translation Learning Collaborative: HCPR - A&P PP Presentation (2015)
4. Tandara, A.A., Mustoe, T.A.: The role of the epidermis in the control of scarring: evidence for mechanism of action for silicone gel. J. Plast. Reconstr. Aesthet. Surg. **61**, 1219–1225 (2008)
5. Groh, M., et al.: Deep learning-aided decision support for diagnosis of skin disease across skin tones. Nat. Med. **30**, 573–583 (2024)

6. Oninla, O.A., et al.: African Skin: Different Types, Needs and Diseases. Int. J. Trop. Dis. Health. 1–13 (2019)
7. Orji, J.U., Chan, G.: Augmented Reality and Machine Learning in Health: A Systematic Review
8. Chen, L., Day, T., Tang, W., John, N.W.: Recent Developments and Future Challenges in Medical Mixed Reality (2017)
9. Gour, S., et al.: Impact of machine learning techniques in medical treatment assistance in perspective to AR & VR technologies. In: E3S Web of Conferences. EDP Sciences (2024)
10. Akyeramfo-Sam, S., Addo Philip, A., Yeboah, D., Nartey, N.C., Kofi Nti, I.: A web-based skin disease diagnosis using convolutional neural networks. Int. J. Info. Technol. Comp. Sci. **11**, 54–60 (2019)
11. Omotosho, L., Sotonwa, K., Adegoke, B., Oyeniran, O., Oyeniyi, J.: An automated skin disease diagnostic system based on deep learning model. J. Eng. Stud. Res. **27**, 43–50 (2022)
12. Pangti, R., et al.: A machine learning-based, decision support, mobile phone application for diagnosis of common dermatological diseases. J. Eur. Acad. Dermatol. Venereol. **35**, 536–545 (2021)
13. Yotsu, R.R., Ding, Z., Hamm, J., Blanton, R.E.: Deep learning for AI-based diagnosis of skin-related neglected tropical diseases: a pilot study. PLoS Negl. Trop. Dis. 17 (2023)
14. Yotsu, R.R., et al.: The Early Detection and Case Management of Skin Diseases With an mHealth App (eSkinHealth): Protocol for a Mixed Methods Pilot Study in Côte d'Ivoire. JMIR Res Protoc. 11 (2022)
15. McAuley, E., Duncan, T.: McAuley Duncan and Tammen_Psychometric Properties of IMI in Sport (1989)
16. Schmid, D.: The technology acceptance scale: Its Bayesian psychometrics assessed in a factor analysis via Markov chain Monte Carlo models. Hum Factors Ergon Manuf. **32**, 66–88 (2022)
17. Bangor, A., Kortum, P.T., Miller, J.T.: An empirical evaluation of the system usability scale. Int. J. Hum. Comput. Interact. **24**, 574–594 (2008)
18. Hyzy, M., et al.: System Usability Scale Benchmarking for Digital Health Apps: Meta-analysis. JMIR Mhealth Uhealth. 10
19. Thomas, D.R.: A general inductive approach for analyzing qualitative evaluation data. Am. J. Eval. **27**, 237–246 (2006)
20. Braun, V., Clarke, V.: Using thematic analysis in psychology
21. Nielsen, J.: Why You Only Need to Test with 5 Users (2000)

Learning Through Restoration: Engaging Virtual Technologies in Design

Denys Borysenko(✉) and Wang Lei

Simon Kuznets Kharkiv National University of Economics, Prosp. Nauky, 9A, Kharkiv 61166, Ukraine

denis.borysenko@hneu.net

Abstract. This study investigates the pedagogical potential of virtual restoration in design education through an immersive, semester-long project based on the Dunhuang frescoes. Utilizing the SimLab platform, 24 students engaged in virtual re-creation of ancient murals, combining technical training in 3D modeling with historical interpretation and ethical reflection. Results indicate that immersive restoration significantly enhanced student engagement, developed technical and design competencies, and fostered interdisciplinary collaboration. Students reported a heightened sense of cultural awareness and responsibility, illustrating how virtual environments can serve as powerful integrative spaces between digital tools and heritage studies. The findings highlight the value of culturally grounded virtual experiences in cultivating reflective, ethical, and collaborative design practices. Positioned within the broader trend of digital transformation in education, this project demonstrates how virtual environments can simulate authentic, real-world scenarios to bridge theoretical and practical learning. Immersive interaction with heritage content enabled students to not only analyze but meaningfully participate in the conservation discourse. The 3D restoration tasks, combined with structured reflection, fostered deeper cognitive and emotional engagement than traditional classroom settings. These results contribute to a growing body of evidence supporting virtual reality as an effective medium for interdisciplinary education. The study provides a model for integrating virtual heritage into design pedagogy, emphasizing experiential learning, cross-domain thinking, and the ethical use of AI and digital tools in cultural contexts. It further recommends investment in teacher training, infrastructure development, and curriculum design to sustainably implement such technologies in educational practice.

Keywords: Virtual Technologies · Fresco Restoration · Dunhuang

1 Introduction

1.1 Virtual Shift in Cultural Heritage Education

In recent years, the integration of virtual technologies into education has reshaped how students engage with learning environments, offering new modalities for experiential, interdisciplinary, and collaborative knowledge-building. In particular, virtual

M. E. Auer et al. (Eds.): IMCL 2025, LNNS 1954, pp. 13–22, 2026.
https://doi.org/10.1007/978-3-032-24998-2_2

reality (VR), 3D modeling, and interactive digital environments have gained attention in design education, enabling students to navigate complex systems, simulate real-world tasks, and engage in meaningful learning beyond traditional classroom settings [3, 5].

One emerging application of these technologies is within the field of cultural heritage. As museums, archives, and academic institutions increasingly digitize historical artifacts, virtual restoration has evolved from a technical challenge into a pedagogical opportunity. Instead of merely preserving cultural objects, immersive restoration platforms can invite students to critically engage with the past—analyzing, interpreting, and even reconstructing cultural artifacts in simulated environments [1]. This shift allows for the development of not only technical skills, but also cultural empathy and ethical reflection, especially when learners participate in restoring artworks with historical and symbolic weight.

1.2 Pedagogical Applications of Virtual Restoration in Cultural Heritage

The Dunhuang frescoes—ancient Buddhist murals located along the Silk Road—offer a compelling case for such integration. These paintings are widely regarded as masterpieces of religious, historical, and aesthetic value. However, their physical fragility makes access and restoration particularly difficult. Leveraging digital tools like the SimLab platform, educators can recreate these frescoes in immersive 3D environments, enabling students to participate in the process of virtual restoration.

While prior research has examined the use of VR in cultural heritage contexts [2, 4], little attention has been paid to how such tools can be used systematically in design education—not only to teach technical modeling, but to instill historical awareness and ethical sensibility. This study addresses that gap by investigating how students interact with digital heritage materials through a semester-long virtual restoration project, and how such engagement contributes to their learning and motivation.

This research focuses on three guiding questions:

- How does virtual restoration influence students' engagement, learning processes, and motivation in design education?
- What kinds of technical, reflective, and ethical competencies can students develop through immersive interaction with cultural heritage materials?
- Can virtual platforms serve as effective integrative spaces between digital design and heritage studies?

By exploring these questions, the study seeks to contribute to the growing discourse on immersive learning, heritage pedagogy, and interdisciplinary curriculum design in the digital age.

2 Literature Review

2.1 Virtual Reality in Higher Education

Virtual reality (VR) technologies have become increasingly prominent in higher education due to their capacity to simulate authentic, complex learning environments. These environments allow students to engage in experiential, constructivist, and situated learning by enabling them to visualize, manipulate, and interact with digital spaces (Dalgarno & Lee, 2010; Fowler, 2015). In particular, VR offers unique opportunities for embodied learning through spatial representation, immersive narrative, and real-time feedback, which are particularly useful for domains like architecture, cultural heritage, and design [5, 7].

In design education, VR supports a shift from passive instruction to active knowledge creation. It fosters technical fluency in spatial modeling, prototyping, and visualization [9], while also enhancing students' creative confidence by allowing them to test and iterate in simulated settings without physical constraints. Huang et al. [8] demonstrated that VR-based environments also enhance learner autonomy, emotional engagement, and flow—three factors closely tied to motivation and knowledge retention in creative disciplines.

However, as Fowler [14] and Parong & Mayer [10] caution, the effective use of VR depends not only on immersive technology but also on pedagogical alignment. When instructional goals, content, and VR tools are mismatched, students may become overwhelmed or distracted by the novelty of the medium. This calls for a shift from technology-centered deployment toward purpose-driven instructional design, where VR is employed as a pedagogical instrument rather than a showcase tool [4].

Moreover, the reflective dimension of learning in virtual environments is often overlooked. Schön's [11] concept of the reflective practitioner is particularly relevant in design contexts, where decision-making involves ambiguity, iteration, and ethical reasoning. Integrating VR with guided reflection—such as through journaling or peer critique—can support students in articulating not only what they designed but why and how they made specific decisions [12].

VR also holds potential to enhance interdisciplinary collaboration. Lattuca et al. [13] argue that authentic, ill-structured tasks presented in simulated settings foster synthesis across domains, especially when learners are prompted to apply both analytical and interpretive skills. In design education, this could mean blending technical competencies (e.g., modeling) with cultural or ethical analysis—precisely the kind of integration required in virtual restoration projects involving cultural heritage [6].

Thus, while VR offers powerful affordances for experiential and creative learning, its true educational potential lies in structured, reflective, and interdisciplinary use—particularly when applied to meaningful content domains such as historical artifacts or real-world design problems.

2.2 Learning through Cultural Heritage and Immersive Experience

The field of digital heritage has evolved significantly over the past two decades—from its initial emphasis on digitization and preservation to more dynamic forms of engagement and interpretation. Early efforts focused on safeguarding artifacts through high-resolution scans, 3D models, and digital archives. However, with the maturation of virtual and augmented reality (VR/AR) technologies, cultural institutions have increasingly shifted toward immersive experiences that simulate historical spaces and narratives [1, 4].

Bekele et al. [1] classify digital heritage applications into three functional categories: documentation, visualization, and interaction. Among these, interaction is increasingly recognized as the most pedagogically impactful, as it allows users not just to view cultural objects but to manipulate, contextualize, and narratively engage with them. This shift supports deeper emotional and cognitive connections—an insight confirmed by Roussou [18], who found that immersive interaction in heritage settings enhances memory retention and personal relevance.

Recent frameworks advocate moving beyond technical accuracy or aesthetic realism toward more critical forms of engagement. Champion [2] introduces the concept of "critical heritage," in which digital environments are not neutral windows to the past, but platforms that foreground questions of authorship, selection, and interpretation. This approach is echoed by Parry [17], who argues that digital heritage should not only represent the past, but encourage users to reflect on how the past is constructed and contested in the present.

Educationally, this shift is significant. When learners engage with heritage through participatory and reflective frameworks, they develop not only historical understanding but also ethical reasoning and cultural sensitivity—key competencies in today's globalized and pluralistic design practices [16]. Projects that allow students to simulate restoration, reinterpret digital archives, or co-curate virtual exhibitions foster what Smith (2006) calls the "performance of heritage"—a mode of engagement where meaning is co-produced, rather than consumed [19].

Moreover, digital heritage has increasingly incorporated co-design and community collaboration. Researchers like Giaccardi [15] argue for a "heritage 2.0" model, where technology enables lay participants—students, citizens, or marginalized groups—to contribute to digital storytelling and re-contextualization of the past. This democratized approach aligns with constructivist learning theories and supports the development of inclusive, plural narratives within education.

As such, the digital heritage domain no longer simply preserves the past—it invites students, designers, and communities to reimagine, critique, and co-create it.

2.3 Virtual Restoration as a Cross-Disciplinary Learning Tool

Virtual restoration—defined as the digital re-creation or simulation of damaged or lost cultural artifacts—offers a unique pedagogical opportunity to merge design, technology, and heritage studies. Rather than positioning students as passive recipients of cultural knowledge, virtual restoration tasks allow them to engage as active interpreters and reconstructors of the past.

This interdisciplinary approach is particularly potent when situated within project-based learning modules that combine technical instruction (e.g., 3D modeling, texture editing) with historical analysis and ethical reflection. It enables students to understand the implications of their design decisions, grapple with conflicting interpretations, and collaborate across disciplinary boundaries. Such integration reflects a broader shift in design education—from mastery of tools to cultivation of reflective, responsible, and culturally informed design thinking.

Despite its promise, few studies have systematically assessed how virtual restoration influences student learning outcomes, motivation, and interdisciplinary competence. This gap underscores the need for pedagogical frameworks that can guide the integration of virtual heritage platforms—such as SimLab—into design curricula.

Preliminary evidence suggests that engaging with virtual restoration fosters a deeper sense of ownership and agency in learners. When students participate in reconstructing culturally significant artifacts, they are not only developing technical and aesthetic sensibilities but also confronting questions about authenticity, preservation ethics, and cultural context. These experiences can lead to a more nuanced understanding of the role designers play in shaping narratives of the past through digital means. The complexity of these tasks encourages learners to critically evaluate the consequences of their interventions, moving beyond surface-level reproduction to thoughtful interpretation.

The collaborative nature of virtual restoration projects reinforces transferable skills such as communication, problem-solving, and interdisciplinary negotiation. Students must often mediate between historically accurate representation, technological limitations, and contemporary design sensibilities. In doing so, they develop the capacity to work within diverse teams, appreciate multiple knowledge systems, and synthesize insights from various domains. These competencies are increasingly vital in professional contexts where design intersects with technology, culture, and public engagement.

3 Methodology

This study adopted a case-based learning approach situated within a semester-long design module that integrated digital heritage and virtual reality technologies. The objective was to explore how immersive digital restoration affects students' learning experiences, motivation, and interdisciplinary competencies in design education.

3.1 Instructional Design and Learning Context

The module was conducted in a university-level design program and involved 24 students with diverse backgrounds in digital media, architecture, and visual arts. The central tool was SimLab Soft, a 3D modeling and interactive learning platform that supports virtual reality integration. The focal task of the module was the digital restoration of the Dunhuang frescoes, selected for their historical complexity and aesthetic richness.

Before beginning technical tasks, students were introduced to the historical, religious, and artistic context of the Dunhuang murals through curated readings, digital exhibitions, and instructor-led lectures. This preparatory phase ensured that students approached the project with cultural awareness and interpretive sensitivity.

3.2 Learning Activities

Learning activities were designed to combine hands-on technical training with critical and reflective engagement. The core components included the following:

Navigating and manipulating a virtual replica of the Dunhuang cave environment using VR headsets and the SimLab interface.

Performing restoration simulations involving color correction, pattern reconstruction, and texture filling.

Participating in structured group discussions on issues such as digital authenticity, ownership, and ethical restoration practices.

Maintaining individual learning journals to document decisions, challenges, and reflections throughout the module.

Thcsc tasks supported the acquisition of 3D modeling skills and fostered the development of cultural empathy and design ethics. The structure of the activities aligned with theories of experiential learning and reflective practice, encouraging students to engage deeply with both the technical and humanistic dimensions of digital heritage work.

3.3 Data Collection

To assess the impact of the learning experience, a multi-method qualitative research design was implemented. Data were collected from four sources:

- Pre- and post-module surveys measuring students' self-assessed competence, motivation, and attitudes toward digital heritage
- Structured interviews with a selected subset of students at the conclusion of the course
- Non-participant observation during virtual lab sessions, with attention to collaboration patterns and problem-solving behavior

Analysis of student-generated digital artifacts, including restored fresco segments and journal reflections.

This triangulation of methods enhanced the validity of the findings and allowed for both individual and collective learning trajectories to be analyzed.

4 Results

The analysis of survey responses, observation notes, interview transcripts, and student-generated materials yielded several key findings. These results are organized under four thematic categories: increased engagement and motivation, technical skill development, enhanced cultural understanding, and interdisciplinary collaboration.

4.1 Engagement and Motivation

Pre- and post-module survey results revealed a marked increase in students' reported interest in cultural heritage topics. Many students indicated that the immersive nature of the virtual environment made them feel more connected to the restoration task, citing the "presence" of being in the simulated Dunhuang cave as a key motivational factor.

These observations align with prior research emphasizing the motivational potential of immersive media in education [5].

In post-course interviews, several participants described how working on a historically meaningful artifact gave them a sense of purpose not often found in abstract studio projects. One student noted that "this project felt more real than any digital design task we've done—it was like time-traveling with responsibility."

4.2 Technical and Design Competencies

Students demonstrated significant improvement in specific technical competencies, especially in texture reconstruction, digital color calibration, and spatial modeling. Through interaction with the SimLab interface, students developed not only tool fluency but also design problem-solving strategies appropriate to real-world restoration scenarios. These findings support existing literature on the affordances of 3D virtual environments for spatial learning and skill acquisition [3].

Digital outputs submitted by students showed a progressive refinement in surface detail, symmetry correction, and cultural referencing over the 12-week period. In journals and critiques, students reflected on how restoration design required them to adapt their usual design process to account for partial data, visual ambiguity, and historical integrity.

4.3 Cultural Understanding and Ethical Reflection

Throughout the course, students increasingly expressed awareness of the ethical dimensions of restoration. Group discussions frequently explored questions around representation, cultural ownership, and the limits of digital re-creation. These conversations were mirrored in students' journals, where they grappled with the tension between artistic interpretation and historical accuracy.

This reflective engagement suggests that immersive restoration can function not only as a technical task but also as an ethical encounter, where students must reconcile modern digital methods with ancient cultural contexts. These results reinforce Champion's [2] call for integrating critical heritage thinking into virtual environments.

4.4 Interdisciplinary Awareness and Collaboration

Finally, the project appeared to foster interdisciplinary thinking. Students with backgrounds in art, architecture, and engineering reported that the project encouraged them to think across disciplinary boundaries. Working with historical content in a design software environment forced students to integrate visual, historical, and technical knowledge in new ways.

Observational notes confirmed that students frequently consulted each other's disciplinary expertise during lab sessions, enhancing collaborative learning. These interactions reflect the value of project-based, cross-disciplinary models in design education, as emphasized by Kolb (1984) and Merriam (2009).

5 Discussion

The findings of this study underscore the pedagogical value of integrating virtual restoration into design education, particularly when framed within immersive, culturally significant contexts such as the Dunhuang frescoes. The results confirm and extend existing research on the educational potential of VR, while also contributing novel insights into how digital heritage tools can foster ethical awareness and interdisciplinary learning.

5.1 Enhancing Engagement Through Immersive Context

The increased motivation and engagement reported by students align with prior research that identifies immersion and presence as key drivers of educational effectiveness in VR environments [3, 5]. However, this study adds nuance by demonstrating that cultural significance further amplifies these effects. Unlike abstract simulations, the Dunhuang context provided students with a historical and emotional anchor, enhancing their sense of relevance and responsibility.

This supports Champion's [2] assertion that virtual heritage environments must do more than replicate—they must invite interpretation and meaning-making. In this case, the simulated restoration task created a reflective space where students engaged not only technically but also affectively with the past.

5.2 Bridging Skill and Reflection in Design Pedagogy

While previous studies have emphasized VR's potential for spatial and technical skill development (Bekele et al., 2018), this study demonstrates that virtual restoration can serve as a dual-purpose tool—both technical and reflective. Students were not simply learning to use SimLab; they were learning to design with awareness, navigating ambiguity, partial information, and ethical uncertainty.

This dual focus echoes Kolb's (1984) experiential learning theory, which argues that concrete experience and reflective observation are equally essential. The restoration task, embedded within a culturally meaningful context, encouraged students to consider the implications of their design choices, rather than merely their aesthetic or functional success.

6 Conclusion

This study highlights the value of virtual restoration as an experiential and interdisciplinary approach to design education. Through immersive engagement with the Dunhuang frescoes via SimLab, students not only developed technical modeling skills but also gained cultural awareness, ethical sensitivity, and collaborative abilities.

Virtual environments enriched student motivation and learning by contextualizing abstract design tasks within historically meaningful settings. Compared to traditional methods, this approach offered deeper engagement and critical reflection.

To advance such pedagogical innovation, institutions should integrate virtual heritage into design curricula, support teacher training in immersive tools, and foster interdisciplinary collaboration. As digital technologies reshape education, culturally grounded virtual learning offers a compelling path toward more ethical, reflective, and engaged design practice.

The findings point to the pedagogical significance of making students active participants in cultural interpretation. By working directly with digital reconstructions of heritage artifacts, learners were empowered to engage not only with form and technique but with the historical narratives and ethical dilemmas embedded in the materials. This shift from passive observation to active co-creation enhances cognitive, emotional, and ethical dimensions of learning.

The implementation of tools such as SimLab also revealed important considerations for curriculum design and technological infrastructure. While student feedback was largely positive, technical challenges—such as processing delays and usability barriers—highlight the need for robust institutional support and continuous software development. Addressing these challenges will be essential for scaling and sustaining immersive practices across educational contexts.

Finally, this study supports the broader argument that virtual restoration represents not only a methodological innovation but also a cultural and ethical one. It opens up new pedagogical spaces where future designers can confront real-world challenges through interdisciplinary inquiry, digital experimentation, and reflective practice. As such, immersive virtual heritage education is not just an enhancement of current curricula—it is a necessary evolution for preparing students to navigate the complex intersections of technology, culture, and responsibility in the 21st century.

References

1. Bekele, M.K., Pierdicca, R., Frontoni, E., Malinverni, E.S., Gain, J.: A survey of augmented, virtual, and mixed reality for cultural heritage. J. Comp. Cultur. Heritage **11**(2), 1–36 (2018)
2. Champion, E.: Critical heritage and virtual environments. In: Cultural heritage and new media, pp. 173–186. Routledge (2015)
3. Dalgarno, B., Lee, M.J.W.: What are the learning affordances of 3-D virtual environments? Br. J. Edu. Technol. **41**(1), 10–32 (2010)
4. Economou, M., Pujol, L.: Educational tool or expensive toy? Evaluating VR evaluation in cultural heritage. Virtual Reality **11**(4), 245–260 (2007)
5. Radianti, J., Majchrzak, T.A., Fromm, J., Wohlgenannt, I.: A systematic review of immersive virtual reality applications for higher education: Design elements, lessons learned, and research agenda. Comput. Educ. **147**, 103778 (2020)
6. Barron, B., Darling-Hammond, L.: Teaching for meaningful learning: A review of research on inquiry-based and cooperative learning. In: Powerful learning: What we know about teaching for understanding, pp. 11–70. Jossey-Bass (2008)
7. Freina, L., Ott, M.: A literature review on immersive virtual reality in education: State of the art and perspectives. eLearning and Software for Education **1**, 133–141 (2015)
8. Huang, H.M., Rauch, U., Liaw, S.S.: Investigating learners' attitudes toward virtual reality learning environments: Based on a constructivist approach. Comput. Educ. **55**(3), 1171–1182 (2010)
9. Lee, E.A.L., Wong, K.W.: Learning with desktop virtual reality: Low spatial ability learners are more positively affected. Comput. Educ. **79**, 49–58 (2014)
10. Parong, J., Mayer, R.E.: Learning science in immersive virtual reality. J. Educ. Psychol. **110**(6), 785–797 (2018)

11. Schön, D.A.: The reflective practitioner: How professionals think in action. Basic Books (1983)
12. Sclater, M.: Theorising from within digital arts-based research: Visualising education futures with young people through co-design. Int. J. Educ. Arts **17**(30), 1–21 (2016)
13. Lattuca, L.R., Voigt, L.J., Fath, K.Q.: Does interdisciplinarity promote learning? Theoretical support and researchable questions. Rev. High. Educ. **28**(1), 23–48 (2004)
14. Fowler, C.: Virtual reality and learning: Where is the pedagogy? Br. J. Edu. Technol. **46**(2), 412–422 (2015)
15. Giaccardi, E.: Heritage and social media: Understanding heritage in a participatory culture. Routledge (2012)
16. Kalay, Y. E., Kvan, T., Affleck, J.: New heritage: New media and cultural heritage. Routledge (2008)
17. Parry, R.: Recoding the museum: Digital heritage and the technologies of change. Routledge (2007)
18. Roussou, M.: Virtual heritage: From the research lab to the broad public. Bar International Series **1075**, 93–100 (2002)
19. Smith, L.: Uses of heritage. Routledge (2006)

AI-Driven Inpainting for 3D Mesh Reconstruction in Virtual Reality: A Systematic Review

Johannes H. Moolman(✉), Fiona Boyle, and Joseph Walsh

Munster Technological University, Dromtacher V92 HD4V, Co. Kerry, Ireland
hans.moolman@mtu.ie

Abstract. Integrating Virtual Reality (VR) into higher education accelerates the need for high-quality 3-Dimensional (3D) assets. While 3D meshes created from 2-Dimensional (2D) images or videos offer an accessible means of content generation, they often suffer from incomplete geometry due to occlusions or limited perspectives. Recent advancements in AI (Artificial Intelligence)-based inpainting techniques have shown promise in addressing these gaps by reconstructing missing mesh data. This systematic review examined literature from 2020 to 2025, focusing on methods suitable for real-time use on mobile VR platforms, such as the Meta Quest 3. Using a structured search across major databases (IEEE Xplore, ACM DL, arXiv, SpringerLink, Scopus), top studies were selected based on relevance, innovation, and feasibility in low-compute environments. Key methods included mesh-based graph neural networks, 2D-to-3D projection pipelines, and diffusion models. The review found that some lightweight AI inpainting techniques are increasingly capable of generating complete 3D models from sparse input data, potentially eliminating the need for high-end computing. The findings suggest a viable path toward scalable, educator-friendly VR asset creation workflows using only standalone headsets, particularly in fields like engineering, healthcare, and heritage studies.

Keywords: 3D Mesh Inpainting · Virtual Reality in Education · AI-Based Reconstruction

1 Introduction

Creating immersive VR experiences relies on the availability of complete, high-quality 3D models, yet traditional techniques like photogrammetry and LiDAR (Light Detection And Ranging) remain too resource-intensive for real-time or mobile use. In response, AI (Artificial Intelligence)-based inpainting has emerged as a promising alternative, capable of reconstructing missing geometry from 3D meshes generated via 2D images or video. Recent work spans diverse strategies: depth-aware inpainting from RGB-D (Red Green Blue-Depth) inputs [1], voxel-based GANs (Generative Adversarial Networks) [2], mesh refinement using GCNs (Graph Convolutional Networks) [3], and self-supervised completion methods that learn directly from the object [4]. Additional innovations include

M. E. Auer et al. (Eds.): IMCL 2025, LNNS 1954, pp. 23–30, 2026.
https://doi.org/10.1007/978-3-032-24998-2_3

2D projection-based inpainting [5] [6] and volumetric scene editing via NeRFs (Neural Radiance Fields) and diffusion models [7] [8]. While these approaches offer improved realism and mesh integrity for VR, many remain unsuitable for deployment on standalone headsets like the Meta Quest 3 due to their computational complexity. Furthermore, there is a limited consensus on metrics that assess their real-time suitability in educational VR. This review systematically analyses AI-driven 3D inpainting research from 2020 - 2025, focusing on approaches viable for lightweight, headset-based use in immersive learning environments. Our contributions are as follows:

- *Systematic Literature Review*: We synthesise research from 2020 - 2025 on AI-driven inpainting methods for 3D mesh reconstruction, emphasising VR-relevant applications.
- *Performance and Feasibility Evaluation*: We assess which methods are most suitable for real-time or mobile VR use, especially in educational settings.
- *Educational Integration Perspective*: We highlight use cases and requirements specific to VR learning environments (e.g., mechanical engineering, heritage studies).

Review Questions:

- What AI-based inpainting methods have been developed for 3D mesh completion from 2D image/video data?
- Which techniques are most suitable for use on mobile VR platforms in terms of computational efficiency and mesh quality?
- How can these methods be integrated into educational workflows to support accessible 3D asset creation?

2 Literature Review

Recent advances in AI-based inpainting techniques have led to a diverse range of strategies for completing 3D meshes, with implications for real-time VR content creation. *Shih et al.* [1] introduced a learning-based method that transforms a single RGB-D image into a multi-layer 3D photo using depth-aware inpainting, enabling motion parallax in immersive VR. *Wang et al.* [2] proposed a two-stage GAN using voxelised inputs for coarse-to-fine mesh hole filling, producing structurally consistent reconstructions ideal for cleaning 3D scans. *Kim et al.* [3] built on SpiralNet + + by developing a graph-convolutional autoencoder with anisotropic filters, achieving detailed mesh completion even in noisy, deformable scans. *Hattori et al.* [4] offered a self-supervised GCN framework that learns shape priors directly from the input mesh, requiring no external datasets - ideal for in-field VR scanning. *Hernandez-Bautista and Melero* [5] introduced a 2D parametrisation method using curvature maps and CNN (Convolutional Neural Networks)-based inpainting, which reprojects results into detailed 3D mesh patches (Fig. 1).

Gisbert et al. [6] addressed cloth mesh inpainting via UV flattening and a UNet model, producing physically plausible completions supported by the ScarfFolds dataset and real-time interactive editing. *Asija et al.* [9] developed a pipeline for converting a single 360° image into a navigable VR environment by pairing depth estimation with panoramic inpainting, enabling one-shot VR scene generation. *Albanis et al.* [10] applied

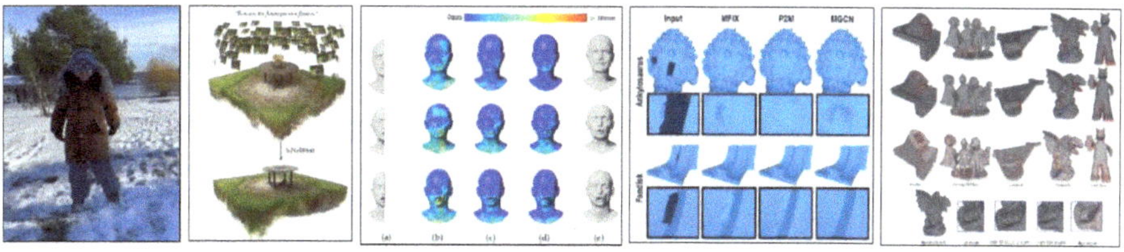

Fig. 1. From L – R: Results from various techniques for AI-Inpainting from *Shih et al.* [1] *Wang et al.* [2] *Kim et al.* [3] *Hattori et al.* [4] *Hernandez-Bautista and Melero* [5]

inpainting for DR (Diminished Reality), removing furniture from 360° images for AR (Augmented Reality)/VR interior design, with super-resolution and structure-guided realism. *Li et al.* [7] presented HHMR (Holistic Hand Mesh Recovery), a diffusion-based framework for completing partial 3D hand meshes with high anatomical plausibility and diversity - essential for VR hand tracking. Finally, *Wang et al.* [8] extended inpainting to NeRFs (Neural Radiance Fields) via InNeRF360, allowing text-guided object removal using NLP (Natural Language Processing), 3D diffusion, and perceptual loss to create editable photorealistic scenes suitable for next-gen VR applications (Fig. 2).

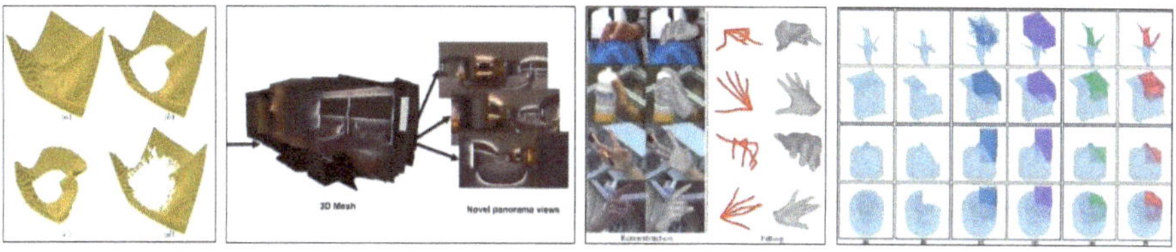

Fig. 2. From L – R: Results from various techniques for Inpainting from *Gisbert et al.* [6] *Asija et al.* [9] *Li et al.* [7] *Wang et al.* [8]

3 Methodology

This systematic review focused on AI-driven inpainting techniques for 3D mesh reconstruction with specific emphasis on VR applications, covering the period from 2020 - 2025. Literature was retrieved from leading databases including IEEE Xplore, ACM Digital Library, SpringerLink, arXiv, Scopus, and Web of Science, using search terms such as "*3D mesh inpainting*", "*deep learning hole filling*", and "*VR 3D reconstruction inpainting*". Filters were applied to select English-language, open-access papers published within the specified timeframe.

3.1 Research Objective and Review Protocol

This review aimed to systematically map, evaluate, and synthesise peer-reviewed literature from 2020 to 2025 on AI-based inpainting techniques that complete missing regions in 3D meshes reconstructed from 2D images or video. The focus was on approaches suitable for real-time VR use, particularly in educational settings. A systematic review methodology was selected to ensure comprehensive, reproducible, and interdisciplinary coverage of advances relevant to immersive learning technologies.

3.2 Search Strategy

The literature search for this review was conducted across six major academic databases: IEEE Xplore, ACM Digital Library, SpringerLink, arXiv, Scopus, and Web of Science. To capture relevant studies, the following Boolean search string was used: "*3D mesh*" AND ("*inpainting*" OR "*hole filling*" OR "*shape completion*") AND ("*AI*" OR "*deep learning*") AND ("*VR*" OR "*virtual reality*"). The search was limited to publications released between January 2020 and March 2025, and in English only, ensuring a contemporary and focused analysis of recent advancements in AI-based inpainting for 3D mesh reconstruction in VR contexts.

3.3 Study Selection

A total of 102 records were identified. After deduplication and screening, 33 full texts were assessed, with 11 papers included in the final synthesis. The selection process is highlighted in Table 1. Below.

Table 1. Inclusion/Exclusion criteria and screening process.

Inclusion criteria	Exclusion criteria	Screening process
Peer-reviewed papers or preprints from top-tier venues	Studies on 2D inpainting without 3D reconstruction	Initial title/abstract review
Explicit focus on AI-based 3D inpainting or shape completion	Methods requiring extensive cloud-based computing	Deduplication deletion
Evaluation on mesh, point cloud, or implicit 3D geometry	Non-English or non-peer-reviewed works	Double screening of 10% of titles/abstracts to ensure reliability
Relevance to VR or general 3D model creation		

3.4 Data Extraction

Data was manually extracted using a structured Excel sheet with the following fields:

- Bibliographic information (author, year, venue)
- Algorithm class (e.g., GAN, GCN, diffusion model)
- Dataset(s) used
- Evaluation metrics (e.g., PSNR (Peak Signal-to-Noise Ratio), CD (Chamfer Distance), EMD (Earth Mover's Distance), perceptual score)
- VR relevance (hardware used, runtime feasibility)
- Code and dataset availability

3.5 Quality and Bias Assessment

The maximum possible score is 14. Studies scoring below 8 were flagged for low methodological robustness but included with caveats. Each included study was assessed using a custom 7-point checklist, scored 0 (absent), 1 (partial), or 2 (complete), as shown Fig. 3. Below.

Author(s)	Reproducibility (code availability)	Dataset Openness	Quantitative Evaluation	VR Context/Application Relevance	Runtime/Performance Reporting	Clarity of Methodology	Benchmarking Against Baselines	Total Score
Shih et al. (2020)	1	1	2	2	2	1	1	**10**
Wang et al. (2020)	2	1	2	2	1	1	2	11
Kim et al. (2022)	1	1	2	1	1	1	1	**8**
Hattori et al. (2024)	1	2	2	2	2	1	2	**12**
Hernandez-Bautista et al. (2023)	1	1	2	1	1	2	2	**10**
Gisbert et al. (2024)	2	2	2	1	2	2	2	13
Asija et al. (2023)	1	2	2	1	2	1	1	**10**
Albanis et al. (2024)	1	2	2	1	0	2	2	**10**
Li et al. (2024)	1	1	1	1	2	2	1	**9**
Wang et al. (2024)	2	2	1	2	0	2	2	**11**
Hernandez-Bautista et al. (2024)	1	1	2	2	2	2	1	**11**

Fig. 3. Quality and bias assessment by publication.

3.6 Data Synthesis

Table 2 below presents a structured overview of the quality and bias assessment, outlining how each study contributed to the qualitative synthesis, quantitative metrics, and subgroup analyses across method type, VR applicability, and input modality.

Table 2. Data synthesis.

Qualitative synthesis	Quantitative synthesis	Subgroup analysis
Method type	Runtime, Accuracy	Mobile vs Desktop
VR applicability	Chamfer Distance avg	GAN vs GCN vs Diffusion
Geometry format	Model size stats	Image vs Depth + Image

4 Results

This section presents an analysis of the final corpus of 11 publications selected through a rigorous systematic review of AI-based inpainting techniques for 3D mesh reconstruction, especially in the context of VR. The studies span from 2020 to early 2025, reflecting a steadily growing interest in leveraging AI for mesh completion tasks suitable for immersive environments. The plot chart shown in Fig. 4. Below illustrates the steady rise in publications on this topic, with notable increases in 2023 and 2024. This trend demonstrates the growing relevance of AI-driven mesh completion, particularly as more mobile-friendly VR solutions (e.g., Meta Quest 3) have emerged.

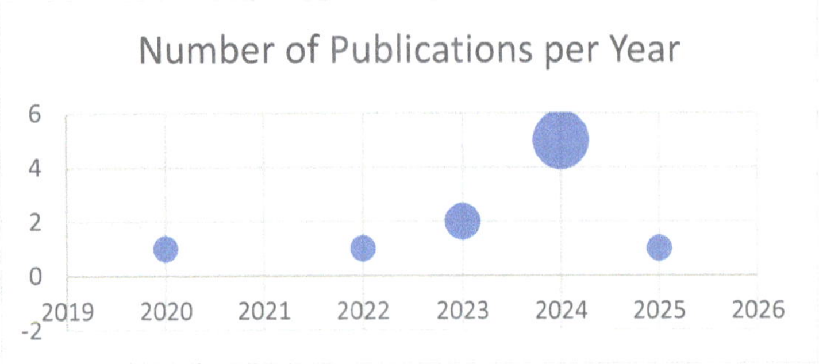

Fig. 4. A plot chart indicating the number of publications by year.

A tabulated summary of the included studies is presented in Table 3, including authorship, publication year, the core technique employed, datasets used, and evaluation metrics. Techniques range from GANs and graph neural networks (GCNs) to self-supervised and hybrid 2D-3D inpainting models, with datasets including ShapeNet, Structured3D, and FAUST. A variety of metrics, such as Chamfer Distance, PSNR, and perceptual quality scores, were used, highlighting a methodological diversity in evaluating mesh inpainting.

Table 3. Study-level summary.

Author	Year	Technique	Dataset	Metric
Shih et al	2020	Layered Depth + Inpainting	Single RGB-D frames	PSNR, Depth Error
Wang et al	2020	3D GAN	ShapeNet	Chamfer Dist
Kim et al	2022	SpiralNet + + GCN	FAUST	Vertex Error
Hattori et al	2024	Self-supervised GCN	Watertight ScanNet	Vertex Delta
Hernandez-Bautista et al	2023	2D Projection + Inpainting	Curvature Maps	Normal Error
Gisbert et al	2024	UV Attention UNet	ScarfFolds	Inpaint Error
Asija et al	2023	Panorama Inpainting	Structured3D	Perceptual Quality
Li et al	2024	Graph Diffusion	FreiHAND	Mesh Accuracy
Albanis et al	2023	Automatic DR-enhanced AR	Structured3D	PSNR, SSIM, MAE, and LPIPS
Hernandez-Bautista et al	2024	Deep learning	Curvature Maps	Normal Error
Wang et al. (InNeRF360)	2024	NeRF Inpainting	Indoor360	LPIPS, SSIM

Table 4 provides a comparative overview of key AI-based inpainting methods, highlighting their input requirements, real-time feasibility, VR readiness, strengths, and limitations across geometry types.

Table 4. Comparative overview of methods.

Method	Geometry type	Real-time capable	VR Ready	Strengths	Limitations
GAN-based	Voxel	No	Medium	Coarse structure learning	Slow, high compute requirements
GCN (SpiralNet + +, etc.)	Mesh	Yes (some)	High	Fine-grained surface detail	Sensitive to noise
Diffusion (Graph)	Mesh	Partial	High	Diverse completions, high fidelity	Requires long inference time
2D Projection + UNet	UV/Curvature	Yes	Medium	Leverages image inpainting strengths	Relies on good parameterization
NeRF Inpainting	Volumetric	No	High	Realistic novel views, scene-level edits	Requires pre-trained NeRF

5 Conclusion

This review synthesised recent research (2020 - 2025) on AI-based inpainting techniques for completing 3D meshes derived from 2D imagery, focusing on enabling immersive educational experiences via standalone VR headsets such as the Meta Quest 3. By analysing 11 representative studies, the review mapped a diverse landscape of approaches - including GANs, GCNs, diffusion models, and hybrid 2D-to-3D projections - each offering different trade-offs between reconstruction quality, computational efficiency, and real-time feasibility.

The findings indicate that graph-based methods, particularly self-supervised GCNs and architectures like SpiralNet + +, offer the best balance of fine-grained surface recovery and runtime performance, positioning them as strong candidates for mobile VR deployment. Meanwhile, diffusion models and GAN-based voxel methods demonstrate high-fidelity outputs but remain too resource-intensive for untethered, real-time use. Hybrid 2D-to-3D projection techniques, which repurpose image inpainting methods, show promise for VR asset repair and editing due to their lower computational demands.

However, a persistent gap lies in the absence of standardised VR-centric benchmarks, especially for perceptual quality and latency, which limit meaningful comparison and practical adoption in educational contexts. Few methods are explicitly optimised for real-time inference or in-headset workflows, underscoring a key opportunity for future research. To advance the field, researchers should focus on lightweight architecture, runtime optimisation, and evaluation in immersive learning contexts. For industry, there is a clear incentive to develop end-to-end pipelines for real-time 3D asset generation and editing directly within standalone VR platforms, reducing reliance on external compute infrastructure and unlocking scalable educational applications.

References

1. Shih, M.-L., Su, S.-Y., Kopf, J., Huang, J.-B.: 3D Photography using Context-aware Layered Depth Inpainting. In: IEEE/CVF Conference on Computer Vision and Pettern Recognition. Seattle (2020)
2. Wang, D., Zhang, T., Abboud, A., Susstrunk, S.: InNeRF360: Text-Guided 3D-Consistent Object Inpainting on 360° Neural Radiance Fields. In: IEEE/CVF Conference on Computer Vision and Pattern Recognition (CVPR). Seattle (2024)
3. Kim, S., Roh, J., Im, H., Kim, J.: Anisotropic SpiralNet for 3D Shape Completion and Denoising. MDPI **22**(17), 6457 (2022)
4. Hattori, S., Yatagawa, T., Ohtake, Y., Suzuki, H.: Learning Self-Prior for Mesh Inpainting Using Self-Supervised Graph Convolutional Networks. IEEE Trans. Visual Comput. Graphics **31**, 1448–1464 (2025)
5. Hernández-Bautista, M., Melero, F.J.: Deep Learning of Curvature Features for Shape Completion. ScienceDeirect **115**, 204–215 (2023)
6. Gisbert, G., Chaine, R., Coeurjolly, D.: Neural Inpainting of Folded Fabrics with Interactive Editing. Computer & Graphics (proc. of Shape Modeling International 2024), vol. 112 (2024)
7. Li, M., et al.: HHMR: Holistic Hand Mesh Recovery by Enhancing the Multimodal Controllability of Graph Diffusion Models. In: IEEE/CVF Conference on Computer Vision and Pattern Recognition (CVPR). Los Alamitos (2024)
8. Wang, X., Xu, D., Gu, F.: 3D Model Inpainting Based on 3D Deep Convolutional Generative Adversarial Network. IEEE Access **8**, 170355–170363 (2020)
9. Asija, S., Du, E., Nguyen, N., Zollmann, S., Ventura, J.: 3D Pano Inpainting: Building a VR Environment from a Single Input Panorama. In: 2024 IEEE Conference on Virtual Reality and 3D User Interfaces Abstracts and Workshops (VRW) (2024)
10. Albanis, G., et al.: An AI-Based System Offering Automatic DR-Enhanced AR for Indoor Scenes. In Advanced Intelligent Virtual Reality Technologies (2023)
11. Hernández-Bautista, M., Melero, F.: 3D Hole Filling using Deep Learning Inpainting. arXiv (2024)

Work-In-Progress: Curating Primary Sources for Civic Themes: VR-Ready Digital Galleries in Teacher Education

Esra Özcan[1(✉)], Tülin Haşlaman[1], and Gökçe Nur Yılmaz[2]

[1] Faculty of Education, Department of Elementary Education, TED University, Ankara, Turkey
esra.ozcan@tedu.edu.tr

[2] Faculty of Engineering, Department of Computer Engineering, TED University, Ankara, Turkey

Abstract. This work-in-progress reports a case study with six groups of preservice teachers (three per group) in a Museum Education course. Each group selects one of the six core themes of the National Council for the Social Studies (NCSS) and curates ten openly accessible artifacts from diverse regions. For each item, students write an about 100-word interpretive label, provide accessibility-oriented alt text, add a source link, and state the license. Galleries are built in Artsteps as 3D exhibitions, shared on the web, and optionally viewed via mobile VR; they are also embedded on the course site. Evidence comes from (i) a group-level curatorial rubric (theme–artifact alignment; selection/sequence/flow; label quality; accessibility and licensing; embedded activities), (ii) interaction logs (clicks, distinct items visited, dwell time, route completion), and (iii) individual reflections. Early expectations are that clearer sequencing/flow and stronger label quality will co-occur with broader coverage and longer dwell times. The study offers a practical framework for integrating digital and VR-ready thematic galleries into pedagogical practice.

Keywords: VR-ready digital galleries · Curatorial practice · Social studies · Teacher education · Learning analytics

1 Introduction

In teacher education, authentic tasks gain meaning when they bring together evidence-based justification, multiple perspectives, and disciplined communication. Within this frame, building curatorial and thematic digital galleries constitutes a concrete enactment of the teacher–curator approach: theme formulation, artifact selection and sequencing, and interpretive label writing are integrated within an online or 3D exhibition environment [1, 2]. The literature indicates that social and digital curation supports authentic learning and contemporary competencies among preservice teachers; however, practice often reveals a need for explicit guidance on selection criteria, curatorial coherence, and platform literacies [3]. At the same time, virtual and 360° gallery–museum environments can broaden access and participation, provided they are anchored by contextualization and clear pedagogical structuring [4].

M. E. Auer et al. (Eds.): IMCL 2025, LNNS 1954, pp. 31–38, 2026.
https://doi.org/10.1007/978-3-032-24998-2_4

Responding to these needs, this work-in-progress paper reports the design of an ongoing case study situated in a Museum Education course at a teacher education program in Turkey. The design integrates (i) curation from open-access collections with license verification, (ii) accessibility-oriented label and alt-text production, (iii) 3D gallery construction via Artsteps, and (iv) a light interaction-logging setup for traceability. The implementation phase is planned for November 2025. The overall aim is to develop and pilot a descriptive framework for positioning thematic, digital, and VR-ready galleries within preservice teacher education practice.

2 Thematic Museums as Learning Environments

A museum is not merely a repository of objects; it is a curatorially structured, theme-oriented learning environment. Contemporary museum studies show how digital platforms and visitor-centered strategies connect on-site and online experiences while democratizing access and participation [5, 6]. In teacher education, this view intersects with the teacher-as-curator approach: exhibitions are designed through choosing a theme, selecting and sequencing artifacts (flow), and composing interpretive labels aligned with learning goals. Thus, curatorial practice reframes exhibitions from static displays into arenas of knowledge production and pedagogy [7–10].

Within this context, thematic and virtual galleries function as effective learning environments that move beyond display. 3D online galleries (e.g., Artsteps, Kunstmatrix, CoSpaces) activate a praxis cycle of action and reflection and foster dialogic interaction through guided stops, open-ended questions, and self-directed tasks [1, 2]. Pandemic-era reviews of "virtual museum" initiatives emphasize that platforms such as Google Arts & Culture expand access; however, learning is not sustained without contextualization and pedagogical scaffolding [4]. Interactive and immersive tools used in museums and learning labs (VR/AR, projection-based installations) help transform visitors from passive observers into active participants, strengthening conceptual understanding and visual thinking [11, 12].

For art-related learning, digital approaches foreground cognitive curation, inclusive access, and collaborative knowledge construction across networked platforms. Post-digital debates further expand the exhibition space with distributed and reusable components [13, 14]. Open-access collections [15, 16] and Creative Commons licenses [17] enable the legal reuse of high-resolution visual and 3D assets; providing clear source–license statements and writing alt text are now regarded as essential for ethics and accessibility [15–17].

Design focus of this study. We operationalize VR-ready thematic digital galleries as structured learning environments along three dimensions: (a) curatorial decisions (theme, selection, sequencing, labeling), (b) learning-activity design (pre-/in-gallery prompts, self-directed micro-tasks, peer explanations), and (c) accessible presentation (web-based delivery with optional VR).

3 Social Studies Context

In social studies, museum education offers a powerful environment for evidence-based inquiry, intercultural comparison, temporal–spatial reasoning, and citizenship-oriented discussion through authentic artifacts. Visitor-centered approaches and digital configurations broaden access and participation; nevertheless, sustainable learning still depends on contextualization and pedagogical structure [4–6].

Within teacher education, the teacher-curator approach integrates thematic formulation, selection and sequencing (flow), and labeling/alt-text into a display structure aligned with instructional purposes [1, 2, 7]. To ensure curricular coherence, this study adopts the National Council for the Social Studies (NCSS) core themes [18] as a guiding framework. Accordingly, artifact selection, sequencing, and labeling are organized through a shared language, while evaluation (via a curatorial rubric) and individual reflections help maintain contextual consistency and make learning processes visible [8, 9, 19].

4 Related Works and Rationale

Studies with preservice teachers indicate that digital and virtual exhibition settings enhance candidates' pedagogical confidence, visual literacy, and self-regulation; however, they also reveal a persistent need for explicit guidelines concerning artifact selection, curatorial consistency, and digital platform literacy [3, 9, 10, 19, 20]. In the context of social studies, central goals such as evidence-based inquiry, intercultural comparison, and temporal–spatial positioning become visible through artifact selection–sequencing and labeling/alt-text practices. The teacher-curator approach thus provides a systematic alignment between instructional objectives and curatorial processes [1, 2, 7].

Empirical and review studies on virtual and 3D galleries highlight that online configurations broaden access and participation, yet when contextualization and instructional scaffolding are lacking, learning continuity weakens [4–6]. Indicators of spatial presence and engagement in 360°/VR environments tend to improve when short practice sessions and coherent waypoint structures are included, underscoring the importance of managing cognitive load [10, 19, 21]. Likewise, findings from museum and gallery-based learning suggest that interactive and immersive tools transform learners from passive spectators into active participants, enhancing conceptual understanding and visual thinking [11, 12].

Ethical and accessibility dimensions have become increasingly significant in curatorial practice. Open-access collections [15, 16] and Creative Commons licensing [17] enable the lawful reuse of high-resolution images and 3D assets. Practices such as including source–license statements and producing alt-text are now considered essential for both ethical integrity and accessibility [15–17]. Literature in art education further emphasizes cognitive curation, inclusive access, and collaborative knowledge construction, while post-digital debates highlight how exhibitions can be expanded through reusable and distributed elements [13, 14].

Taken together, this body of work demonstrates that thematic exhibition design in social studies provides a strong foundation for:

(a) aligning artifact selection and sequencing with learning objectives [1, 9, 10];

(b) fostering evidence-based expression and dialogic engagement through labeling and alt-text [7, 8, 13]; and
(c) expanding access through web-based and VR-ready formats [4–6].

Building on these insights, the present study—situated in a Museum Education course—adopts the NCSS [18] core themes as the guiding framework for VR-ready thematic digital galleries and integrates evidence from three complementary sources: (i) a group-level curatorial rubric (theme–artifact alignment, sequencing/flow, label quality, accessibility/licensing, embedded activities), (ii) visitor interaction logs (click counts, number of distinct artifacts visited, dwell time, route completion), and (iii) individual reflection texts [2, 11, 12]. Prior work indicates that such galleries hold strong potential as structured learning environments in teacher education, while their effectiveness depends on curatorial coherence, accessibility, and structured assessment mechanisms.

Accordingly, this study is guided by the following research questions:

RQ1: What is the level of preservice teachers' group-level curatorial competence, as evidenced by rubric results?

RQ2: How are group-level curatorial rubric scores associated with visitor engagement indicators (clicks, distinct-item coverage, dwell time, route completion)?

RQ3: How do preservice teachers describe their experiences of designing thematic digital/virtual galleries, as revealed through inductive content analysis of individual reflections?

5 Method

5.1 Research Design and Approach

This work-in-progress study adopts a case study design to explore the implementation of VR-ready thematic digital galleries in a Museum Education course. A case study approach is well suited for examining educational practices that unfold as multi-component processes within authentic contexts and through multiple sources of evidence [22]. By situating the gallery-design activity in its natural instructional environment, the study aims to capture both the collective products of preservice teachers and the interactional patterns that emerge during the process.

5.2 Participants and Setting

Participants are preservice teachers enrolled in a Museum Education course at a private university in Turkey. The class is organized into six groups of three. Each group adopts one of the NCSS [18] core themes—Culture; Time, Continuity & Change; People, Places & Environments; Individuals, Groups & Institutions; Power, Authority & Governance; or Global Connections—and curates a thematic digital gallery structured into subthemes.

Galleries are built in Artsteps as 3D environments, published for web-based navigation and optional mobile-VR viewing, and embedded on the course website. Each group documents ten open-access artifacts aligned with its theme, recording metadata (title, maker/culture, date, provenance, material, source URL, license). For every artifact, members prepare an approximately 100-word interpretive label and an alternative text for accessibility; a prompt question is added at the end of the exhibition to stimulate post-visit discussion.

Institutional ethics approval has been obtained; participation is voluntary, and data collection complies with institutional and national regulations. All materials are open-access, accompanied by explicit source and license statements, and alt-text authoring is required.

5.3 Instruments

Three complementary data sources are planned:

- Group-level curatorial rubric: assesses theme–artifact alignment, sequencing/flow, label quality, accessibility & licensing, and embedded learning activities.
- Visitor interaction logs: extracted from platform analytics to capture click counts, number of distinct artifacts visited, dwell time, and route completion.
- Individual reflection texts: short narratives in which preservice teachers describe decisions about selection, sequencing, labeling, accessibility, and collaboration.

5.4 Data Analysis

Quantitative indicators (rubric scores and interaction metrics) will be summarized descriptively. Tables will report means, standard deviations, and, where appropriate, medians and interquartile ranges. Visual representations such as scatter-style summaries will illustrate cross-group patterns, with emphasis on pattern description rather than inference.

Qualitative data from reflection texts will be analyzed using inductive content analysis: (i) familiarization, (ii) open coding, (iii) grouping codes into categories, (iv) reviewing and naming categories, and (v) developing a codebook with illustrative excerpts [23–25]. Two researchers will code independently, with inter-rater reliability computed using Cohen's κ [26].

5.5 Implementation

A four-week preparatory sequence precedes data collection:

1. Technology onboarding: orientation to Artsteps (gallery creation, room design, route setup, VR mode).
2. Theme selection: mapping topics to NCSS cores through exemplar cases.
3. Curatorship workflow: modeling the sequence theme → artifact selection → sequencing → labeling → embedding activities, with sample critiques.
4. Open licensing & accessibility: practicing sourcing, license notation, alt-text writing, and usability checks.

The implementation phase is scheduled for November–December 2025.

6 Expected Findings

Based on the planned implementation, group-level curatorial rubric scores are expected to cluster at moderate-to-high levels, with stronger results in theme–artifact alignment and sequencing/flow and greater variability in accessibility & licensing and label quality.

Higher rubric performance particularly on sequencing/flow and label quality is anticipated to correspond with stronger visitor-engagement patterns such as increased distinct-artifact visits, longer dwell times, and higher route completion rates.

Qualitative reflections are expected to highlight themes related to decision-making in theme–artifact matching, layout and navigation design, clarity of interpretive writing, source–license verification, time and platform constraints, and collaboration. These insights will contextualize variation across groups and inform refinement of the curatorial framework for future iterations.

7 Discussion and Implications

This work-in-progress outlines an instructional arrangement in which preservice teachers enact the teacher–curator role from start to finish formulating a theme, selecting artifacts, organizing sequencing and flow, authoring interpretive labels, and embedding learning-oriented elements into a collective exhibition. The use of a group-level curatorial rubric provides structured feedback on strengths and areas for development, particularly regarding the coherence of theme–artifact alignment, clarity of textual labels, and accessibility and licensing practices.

Within the field of social studies, the NCSS [18] College, Career, and Civic Life (C3) Framework (2010; rev. 2017) serves as a guiding structure that helps preservice teachers deepen conceptual understanding, make interdisciplinary connections, and link citizenship-related content with concrete exemplars. Thematic digital galleries, populated with open-access artifacts from varied regions and historical periods, create a learning environment in which students can identify relationships within a theme and contextualize abstract concepts through visual evidence.

Requiring a source link, license statement, and alternative text for each artifact embeds the ethical–legal and accessibility dimensions of content selection directly into pedagogy. Over time, these practices can cultivate transferable habits of ethical curation and inclusive design for classroom material development. Collectively, these features define a practical pathway for theme-based instruction in social studies, supported by digital and VR-ready exhibition formats.

From a technology integration perspective, 3D builds developed in Artsteps offer an accessible, low-cost approach to immersive learning. The web-based galleries remain fully navigable without VR hardware, while the optional mobile-VR mode adds spatial immersion without compromising route design, readability, or accessibility principles. Publishing the galleries via link and embedding them in the course site supports sustainability and institutional scalability, enabling cross-visits, micro-task sharing, and interdepartmental collaboration without imposing specialized infrastructure.

8 Conclusion

Situated in a Museum Education course, this work-in-progress outlines a framework for VR-ready thematic digital galleries organized at the group level and guided by the NCSS core themes [18]. The framework integrates three complementary evidence sources: (i) a group-level curatorial rubric assessing theme–artifact alignment, selection/sequence/flow, label quality, accessibility & licensing, and embedded learning elements; (ii) visitor interaction logs capturing clicks, distinct-item coverage, dwell time, and route completion; and (iii) individual reflection texts that surface process experiences. Together, these components specify concrete steps for designing thematic digital galleries as learning environments not mere displays where social-studies concepts are examined in context. Embedding open-license and accessibility practices (source link, license statement, alt text) aims to cultivate ethical–legal awareness and inclusive design habits for preservice teachers. A web-first presentation with optional mobile VR offers an adaptable pathway under varied access conditions.

As a single-site case with six groups, this study does not claim generalizability; however, the documented workflow artifact sourcing and documentation, gallery building, label/alt-text authoring, license verification, embedded tasks, and measurement flow, provides a replicable roadmap for similar courses. Future work can extend the study across multiple institutions, employ richer platform analytics, and, where appropriate, compare alternative exhibition layouts to examine interaction patterns and learning outcomes more fully. In this direction, the potential of thematic digital galleries to support theme-based instruction in social studies aligned with open-data and accessibility principles can be investigated with broader samples.

References

1. Tam, C.O.: Teachers Curating Virtual Exhibitions for Learning Visual Arts. J. Art & Design Educ. **42**(3), 514–528 (2023). https://doi.org/10.1111/jade.12471
2. Parsons, A.: Virtual Art Galleries as Learning Spaces and Agents of Praxis. AI, Comp. Sci. Robo. Technol. **2023**(2), 1–23 (2023). https://doi.org/10.5772/acrt.14
3. Gadot, R., Forkosh-Baruch, A.: Preservice Teachers' Practices in Social Digital Curation: Towards Authentic Learning. In: EdMedia + Innovate Learning (Proc.), pp. 942–949. Association for the Advancement of Computing in Education (AACE), Amsterdam, Netherlands (2018). Available at LearnTechLib
4. Verde, A., Valero, J.M.: Virtual Museums and Google Arts & Culture: Alternatives to the Face-to-Face Visit to Experience Art. Int. J. Educ. Res. **9**(2), 49–62 (2021)
5. Giannini, T., Bowen, J.P. (eds.) Museums and Digital Culture: New Perspectives and Research. Springer, Cham (2019). https://doi.org/10.1007/978-3-319-97457-6
6. Bell, D.R., Smith, J.K.: Inside the Digital Learning Laboratory: New Directions in Museum Education. Curator: The Museum Journal **63**(3), 371–386 (2020). https://doi.org/10.1111/cura.12378
7. Lund Dean, K., Forray, J.M.: Educators as Curators: Displaying the Caring Art of Teaching in Management Education. J. Manage. Educ. **44**(6), 739–748 (2020). https://doi.org/10.1177/1052562920945348
8. Rito, C., et al. (eds.): Institution as Praxis: New Curatorial Directions for Collaborative Research. Sternberg Press, Berlin (2020)

9. Sawyer, R.K., Johnson, S.A., Holschuh, M.: Developing teachers as critical curators: investigating elementary preservice teachers' inspirations for lesson planning. J. Teach. Educ. **71**(4), 518–531 (2019). https://doi.org/10.1177/0022487119879894
10. Schroeder, S., Curcio, R.: Critiquing, curating, and adapting: cultivating 21st-century critical curriculum literacy with teacher candidates. J. Teach. Educ. **73**(2), 129–144 (2022). https://doi.org/10.1177/00224871221075279
11. Roldán, J., Lara-Osuna, R., González-Torre, A.: The Project 'Art for Learning Art' in Contemporary Art Museums. J. Art & Design Educ. **38**(3), 566–579 (2019). https://doi.org/10.1111/jade.12245
12. Szubielska, M., Imbir, K.: The Aesthetic Experience of Contemporary Installations in an Art Gallery and a Laboratory Setting: The Issue of Interactivity. Adv. Cogn. Psychol. **18**(3), 179–189 (2022). https://doi.org/10.5709/acp-0360-7
13. Heaton, R.: Digital Art Pedagogy in the United Kingdom. In: The International Encyclopedia of Art and Design Education, vol. 3, pp. 1–19. Wiley (2019). https://doi.org/10.1002/9781118978061.ead003
14. Tavin, K., Kolb, G., Tervo, J. (eds.) Post-Digital, Post-Internet Art and Education: The Future Is All-Over. Palgrave Macmillan, Cham (2021). https://doi.org/10.1007/978-3-030-73770-2
15. Smithsonian Institution: Guide to Interpretive Writing for Exhibitions. Smithsonian Exhibits, Washington, DC (2021). https://exhibits.si.edu/wp-content/uploads/2021/09/SI-Guide-to-Interpretive-Writing-for-Exhibitions.pdf. Accessed 28 Sep 2025
16. The Cleveland Museum of Art: Open Access. (2019–2023). https://www.clevelandart.org/open-access/. Accessed 28 Sep 2025
17. Creative Commons: About CC Licenses (2025). https://creativecommons.org/share-your-work/cclicenses/. Accessed 28 Sep 2025
18. National Council for the Social Studies (NCSS): National Curriculum Standards for Social Studies: Chapter 2—The Themes of Social Studies. NCSS, Silver Spring, MD (2010). https://www.socialstudies.org/national-curriculum-standards-social-studies-chapter-2-themes-social-studies. Accessed 28 Sep 2025
19. Schroeder, S., Siegle, J., Craig, S.D.: Curriculum in the Digital Age: The Teacher-to-Teacher Online Marketplace of Ideas (TOMI). J. Curric. Stud. **56**(5), 683–705 (2024). https://doi.org/10.1080/15505170.2024.2312118
20. Kayaalp, F., Başcı Namlı, Z., Meral, E.: My Museum: A Study of Pre-Service Social Studies Teachers' Experience in Designing Virtual Museums. Educ. Info. Technol. **29**, 24047–24085 (2024). https://doi.org/10.1007/s10639-024-12742-8
21. Fang, F., Jiang, X.: The Analysis of Artificial Intelligence Digital Technology in Art Education Under the Internet of Things. IEEE Access **12**, 22928–22937 (2024). https://doi.org/10.1109/ACCESS.2024.3363655
22. Yin, R.K.: Case Study Research and Applications: Design and Methods, 6th ed. SAGE, Thousand Oaks, CA (2018)
23. Elo, S., Kyngäs, H.: The Qualitative Content Analysis Process. J. Adv. Nurs. **62**(1), 107–115 (2008). https://doi.org/10.1111/j.1365-2648.2007.04569.x
24. Hsieh, H.-F., Shannon, S.E.: Three approaches to qualitative content analysis. Qual. Health Res. **15**(9), 1277–1288 (2005). https://doi.org/10.1177/1049732305276687
25. Schreier, M.: Qualitative Content Analysis in Practice. SAGE, London (2012)
26. Cohen, J.: A coefficient of agreement for nominal scales. Educ. Psychol. Measur. **20**(1), 37–46 (1960). https://doi.org/10.1177/001316446002000104

Building AI Resilience in Education

Building AI Resilience in Engineering Students Using AI Tools

Varghese Panthalookaran(✉) and Rinku Jacob

Rajagiri School of Engineering and Technology, Ernakulam, Kerala, India
panthalookarancmi@gmail.com, rinkuj@rajagiritech.edu.in

Abstract. The technological revolution driven by Artificial Intelligence (AI) is set to fundamentally transform education, particularly in engineering. In the so-called "Age of Intelligence," where intelligence becomes a service akin to utilities like water and electricity, educational paradigms must adapt to remain relevant. While many educators have embraced AI tools to enhance the efficiency of traditional academic processes, they often overlook the crucial need to develop AI resilience in students. This oversight risks a future where engineering and technology graduates become mere users of AI, rendering their roles increasingly redundant in professional settings. An alternative approach focuses on cultivating AI resilience among these AI-native learners. This involves equipping students not only to collaborate effectively with AI tools but also to complement AI's cognitive capabilities with their own distinctive core competencies. This paper presents findings from a series of focused experiments conducted in India since 2021, aimed at fostering AI resilience in engineering students. These ongoing experiments emphasize the development of perceptive and critical thinking skills that are both AI-collaborative and AI-complementary. Participants are tasked with leveraging generative AI tools to build their AI resilience, both individually and in teams. A key component of this process involves students using AI-generated code to simulate physical phenomena underlying laboratory experiments. They are then required to validate the results by comparing the simulated outcomes with actual experimental measurements. This approach prepares students for AI-driven workspaces, where AI tools will handle code development, leaving human professionals to creatively interpret and correlate physical phenomena with their observations. In essence, the pedagogical experiments described in this paper provide a structured framework for equipping AI-native students with the skills necessary to thrive in future work environments by building comprehensive AI resilience.

Keywords: AI resilience · AI Natives · Age of Intelligence · Engineering Education · AI Collaboration · AI Complementarity

1 Introduction

In today's world, artificial intelligence (AI), specifically Generative AI (gen-AI), acts as a powerful extension of human abilities, serving as a collaborative partner in intellectual processes, especially in educational contexts, making education more accessible,

M. E. Auer et al. (Eds.): IMCL 2025, LNNS 1954, pp. 41–58, 2026.
https://doi.org/10.1007/978-3-032-24998-2_5

creative, and efficient [1]. AI tools are designed to work alongside human cognition, enhancing learning and problem-solving in complex tasks. Yet, concerns about Artificial General Intelligence (AGI) remain, as some worry that increasing reliance on AI could diminish critical thinking and potentially challenge human roles in an AI-dominated future [2]. This potential shift necessitates a comprehensive reimagining of education, particularly in engineering, where students are directly involved in the development and application of AI technologies [3]. Students of AI age, the so-called AI natives are expected to master the AI tools and at the same time preserve their prowess for critical thinking [4].

Researchers have extensively studied how integrating AI into engineering education can enhance learning outcomes. For instance, some studies highlight how chatbots can personalize learning experiences, boosting both student engagement and comprehension, essential components of a modern curriculum [5]. Specific AI applications in various engineering disciplines further demonstrate the role of AI in preparing students for industry demands. For example, a study on project-based learning in power generation illustrates how AI-based classifiers are used to diagnose wind turbine faults, giving students hands-on experience in problem-solving and technical skills applicable to industry [6]. In fields such as architecture, engineering, and construction, generative AI fosters adaptive learning and project management skills essential for future professionals in these fields [7]. In civil engineering, AI based tools deepen students understanding of real-world construction projects and the mechanical equipment, preparing them for industry challenges [8]. Similarly, ChatGPT supports students in electrical engineering design courses, helping them tackle complex tasks, fostering industry – relevant skills in AI-assisted problem solving [9]. Additionally, the Portable Experimenter Engineering Board (PEEB) uses AI-supported, hands-on learning to teach engineering students essential skills in programming and electronics, preparing them with industry-relevant knowledge [10].

To ensure students unique contributions in future AI-driven workplaces, they must learn not only to work alongside AI tools but also to cultivate uniquely human strengths that complement AI. This concept, known as "AI resilience," includes two key components: (1) enabling students to collaborate effectively with AI to enhance their cognitive abilities and (2) equipping them with distinctly human skills that AI cannot replicate [11]. Expanding on this, AI-integrated learning environments support "AI resilience" by identifying individual strengths and providing tailored educational experiences. Such environments foster adaptability, resilience, and confidence in students, empowering them to pursue a wide range of talents and interdisciplinary interests, preparing them for meaningful contributions across both traditional and emerging fields [12].

AI resilience encompasses the capabilities required to use AI effectively and ethically in dynamic environments, underscoring the need for educational approaches that prepare students not only technologically but also ethically. Jeon et al. [13] emphasize the critical role of ethical education in fostering AI resilience, arguing that students must be equipped not only with the skills to use AI but also with the capacity to navigate the ethical complexities introduced by AI technologies. Similarly, Khan et al. [14] explore the role of AI-assisted assessment in evaluating young learners' vocabulary knowledge, examining the influence of parental and teacher support on building immunity and resilience. Their study highlights the potential of AI to enhance both the cognitive and emotional dimensions of learning, emphasizing the importance of a supportive educational environment that actively involves parents and teachers in the learning process.

Together, these studies illuminate the multifaceted nature of AI resilience in education, addressing its technical, ethical, and emotional dimensions. However, while various aspects of AI resilience have been examined in relation to education, a comprehensive understanding of how AI tools can be leveraged to achieve AI resilience in engineering education remains underexplored. This paper seeks to address this gap and investigates the specific AI-collaborative and AI-complementary skills that engineering students must cultivate to stay relevant in the era of intelligent machines. As outlined in the pedagogical experiments, students are encouraged to develop two foundational intellectual skills—perceptive and thinking skills— that are both AI-collaborative and AI-complementary [15]. Cognitive learning theory [16] suggests that perceptive and thinking skills evolve through assimilation and accommodation, contributing to individual schemas, which encompass an individual's mindset (cognitive), value-set (affective), skill set (psychomotor), and life-set (sustainability)—along with other competencies that learners develop progressively. By structuring engineering education to foster "AI resilience", students will gain the AI-collaborative skills, such as data perception and scenario creation, as well as the AI-complementary skills, like socio-emotional and aesthetic perception, essential for thriving in AI-augmented workplaces.

In summary, education for AI natives in general, and engineering education in particular, requires reimagining to ensure it is both AI-collaborative and AI-complementary. Together, these needs define a new educational dimension called "AI resilience.". This paper aims to comprehensively delineate the aspects of AI resilience in engineering education, specifically exploring how engineering students can acquire AI-native perceptive skills, thinking skills, and simulation skills. A blueprint for AI-resilient education is available in [17, 18].

2 Suitable Pedagogy for the AI-Natives

This paper emphasizes pedagogical solutions to the challenges AI natives face in their intellectual development. It explores the search for suitable pedagogical approaches for today's students—often called AI natives—who need to become AI-resilient in both life and work environments to thrive. A pedagogy for digital natives is outlined in [19], emphasizing the development of an entrepreneurial mindset in students. This pedagogy enables them to navigate the increased Volatility, Uncertainty, Complexity, and Ambiguity (VUCA) of workplaces affected by AI integration [20].

The entrepreneurial pedagogy for AI natives reimagines education as a process fueled by intrinsic learning motivations and oriented toward the creation of entrepreneurial learning outcomes. This pedagogy identifies five powerful intrinsic motivations for learning: 1) the joy of knowledge creation, 2) the joy of self-actualization, 3) the joy of self-expression, 4) the joy of playful synergy, and 5) the joy of selfless service. Entrepreneurial pedagogy aims to enhance these intrinsic motivations so that each idea presented to a student, results in a productive learning outcome. In other words, this pedagogy enables students to develop a startup idea from a concept by leveraging AI tools.

As a result, students gradually develop an entrepreneurial mindset, which is essential for navigating the VUCA environment introduced by AI tools. An entrepreneurial mind adopts autonomous learning style that promotes development of vision that counters volatility, lifelong learning style that deepen their understanding to surpass uncertainty, venturesome learning style that instils courage or audacity that manages complexity, and pursue flexible learning that improves adaptability to compensate for ambiguity of a VUCA-driven world [19].

Developing an entrepreneurial outcome from an idea with the support of AI may not be overly challenging for AI natives. However, doing so in an AI-collaborative and AI-complementary manner, effectively employing both human and AI core competencies, presents a real challenge. Addressing this goal while making optimal use of AI tools further intensifies the challenge. This paper reports on ongoing educational experiments with the Engineering Physics course, aimed at addressing these concerns [21].

3 The Curricular Process

The curricular process for developing entrepreneurial insight from a concept in physics involves a comprehensive sequence of steps. This structured approach is visualized in Fig. 1, which outlines the key stages from the initial ideation to the final assessment. The process includes: A) Harvest of Seed Ideas, B) AI-Native Elaboration, C) Development of AI-Native Perceptive Skills, D) AI-Native Experimentation, E) AI-Native Simulation, F) Development of AI-Native Thinking Skills, G) Entrepreneurial Sprouting, H) Entrepreneurial Skill Development, I) Synergizing of the Startup Idea, J) Simplification of the Startup Idea, K) Pitching the Startup Idea, and L) Entrepreneurial Assessment. Each of these steps will be discussed in detail below.

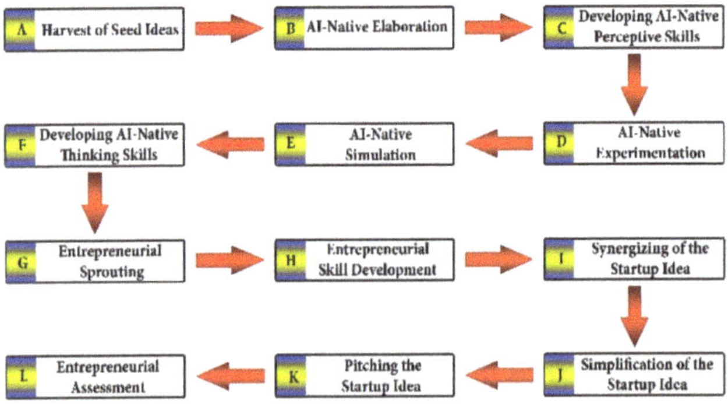

Fig. 1. Flowchart illustrating the progressive stages in the curricular process for cultivating entrepreneurial insight from physics concepts.

3.1 Harvest of Seed Ideas

The first stage of the curricular process focuses on the harvesting of seed ideas, which are core concepts distilled from the syllabus, intended to foster an entrepreneurial mindset in students and inspire them to create entrepreneurial learning outcomes. These are 'ideas in a nutshell' that help integrate curricular, co-curricular, and extracurricular activities into a unified curriculum focused on entrepreneurial education. A universal framework for entrepreneurial curriculum is presented in [22].

In this approach, the entire range of physics topics in the syllabus is condensed into a set of summary concepts referred to as "seed ideas" or "idea eggs." Each seed idea is crafted to encapsulate a distinct physical law or phenomenon that has historically contributed to technological advancement. Collectively, the seed ideas cover the entire syllabus or the full scope of learning content for first-year engineering students. In this experiment, eighty seed ideas were identified from the syllabus, and each student was assigned one unique idea for independent exploration and reflection.

Students are asked to visualize the seed idea as a fundamental law of nature, a principle that has already proven its worth as the foundation of numerous technologies and solutions across diverse fields of human endeavour. Each seed idea has historically inspired entrepreneurial movements that have driven progress, transforming the planet and its people. Each student is encouraged to think like those visionary pioneers who have worked with similar ideas to generate impactful entrepreneurial outcomes. With this goal in mind, students are guided to deeply consider these ideas until they 'hatch' or germinate into entrepreneurial insights. Throughout this reflective process, AI tools support students by building both AI-collaborative and AI-complementary skills.

Further, these seed ideas are tested in a specially designed laboratory, where every harvested seed idea from the syllabus can be experimentally explored. Rooting the entire process in seed ideas drawn from the curriculum ensures that the development of entrepreneurial mindsets is a purely curricular process, thus avoiding the need to relegate it to co-curricular, extracurricular, or intermural activities that would add to the workload of students and teachers.

AI tools can be effectively employed in harvesting seed ideas from a given syllabus. Each seed idea is crafted as a focused statement that clearly captures various aspects of a physical phenomenon. Different guidelines for designing seed ideas are outlined in [19]. Generative AI tools can facilitate the extraction of seed ideas from learning materials or syllabi.

3.2 AI-Native Elaboration

In the second stage, students deepen their understanding of seed ideas using AI tools, focusing on three areas:

1) Original Elaboration

This involves exploring the mindset of the discoverer of the law of nature encapsulated in the seed idea and immersively understanding the processes involved in its innovation. Students connect directly with the creative and entrepreneurial thinking of the original inventor, allowing them to explore the styles and approaches that led to the discovery of the law of nature enshrined in the seed idea. They examine the theories that explained the same phenomenon before its discovery, along with the cultural and historical context in which the discovery took place. Anecdotes from the inventor's journey are intended to inspire students, encouraging them to think creatively, innovate, troubleshoot, and take initiative. In this way, the creative and entrepreneurial spirit of the inventor is passed on to the modern-day learner.

2) Historical Elaboration

Historical elaboration entails gaining a clear understanding of the evolution of the seed idea over time. Students trace the entrepreneurial impact of the idea, exploring patents, technologies, and applications that emerged from it. They develop a historical narrative that captures the 'idea timeline'—the moment the law of nature was discovered by a physicist, its subsequent impact, its role in today's world, and its potential future trajectory. Students research the literature using AI tools, identifying how the idea has transformed industries, shaped technological advancements, and inspired further innovations.

Assessment criteria:

1. How imaginatively have students captured the entrepreneurial spirit of the inventor of the seed idea?
2. How professionally do the students develop the story of the seed idea, ensuring its entrepreneurial efficacy is fully realized?
3. How inspiring is the storyline that the students develop, connecting the past, present, and future of the seed idea in transforming the world around them?

3) Interdisciplinary Elaboration

Interdisciplinary elaboration explores the seed idea's significance and applications across various fields, preparing students to develop entrepreneurial insights that benefit society. Students analyze the relevance of their seed idea in diverse disciplines beyond physics, such as other sciences, mathematics, literature, art, music, economics, management, education, business, psychology, and politics, etc. This interdisciplinary approach integrates the seed idea into a broader knowledge space, enabling students to appreciate its multifaceted impact and preparing them to work effectively in multidisciplinary environments.

Assessment criteria:

1. How accurate and insightful is the student's interdisciplinary understanding of the seed idea?
2. How broad is their understanding of the idea's interdisciplinary applications?
3. How effectively has the interdisciplinary elaboration contributed to the student's development of socially impactful insights?

Students are expected to conduct these elaborations by leveraging AI tools and applying effective prompt engineering. Teachers can support this by introducing best practices in prompt engineering, helping students develop comprehensive mastery over their unique seed idea. In this process, AI tools serve as 'virtual teachers' or 'teacher avatars,' with human instructors facilitating by sharing expertise in utilizing AI for in-depth exploration. Through this approach, students learn to use AI as both a thinking assistant and a creative partner.

3.3 Development of AI-Native Perceptive Skills

Human perception is subject to influence by previous knowledge, experiences, and worldview. With the advent of practical AI tools, students now engage with AI systems that also carry inherent limitations, including algorithmic and data biases, stochastic errors, and the potential for 'catastrophic forgetfulness.' Therefore, it is crucial that AI-native students develop a spectrum of holistic perceptive skills, allowing them to attain a comprehensive understanding of reality. Such an understanding is essential for creating sustainable engineering processes and products that can address root causes effectively.

In the AI era, engineering students' perceptive skills must be cultivated such that they can both collaborate with and complement the perceptive capabilities of AI. To this end, we propose two categories of perceptive skills: 1) AI-Collaborative Perceptive Skills: Skills that align with perceptive competencies AI has already achieved, allowing human counterparts to leverage AI's strengths to enhance their own perceptive abilities. 2) AI-Complementary Perceptive Skills: Skills that AI lacks and may never fully develop, which humans must contribute to create a comprehensive and holistic perceptive spectrum.

Modern engineering students, as AI natives, need to cultivate both AI-collaborative and AI-complementary perceptive skills to become resilient, competent engineers equipped for the AI age. Through AI-collaborative perceptive skills, students can work alongside AI tools, utilizing them to sharpen their perception. Meanwhile, AI-complementary perceptive skills enable students to bring unique human insights, ensuring that perception remains holistic and, therefore, sustainable.

4) AI-Collaborative Perceptive Skills

Two main types of AI-collaborative perceptive skills have been identified where AI demonstrates high efficacy:

- Differential Perceptive Skills: These facilitate pattern recognition within data, helping individuals predict trends and scenarios. AI's established strengths in these skills can be harnessed in collaboration with humans.
- Integral Perceptive Skills: These allow the identification of causation and correlation, which are valuable for hypothesis generation, summarization, and understanding a phenomenon's transdisciplinary significance. Here, competencies can be shared between humans and AI, enhancing human perceptive capabilities.

5) AI-Complementary Perceptive Skills

In contrast, AI lacks significant competence in certain perceptive skills, where humans provide essential complementary strengths:

- Empirical Perceptive Skills: These include contextual, psychosocial, and systems perception—skills where human experience and insight are essential.
- Transcendental Perceptive Skills: This category encompasses aletheic (Truth), eudaimonic (Goodness), and aesthetic (Beauty) perceptions, aligning with philosophical, ethical, and aesthetic perceptive abilities. Mastery of these skills enables students to contribute unique human perspectives, conserving the irreplaceable role of human perception.

Students can develop this broad, holistic spectrum of perceptive skills by utilizing AI tools, like ChatGPT, as learning aids. Through prompt engineering and selecting targeted exercises, students receive guidance for honing both AI-collaborative and AI-complementary perceptive skills.

Assessment criteria:

1. How effectively does the student develop AI-collaborative perceptive skills using the seed idea?
2. How effectively does the student cultivate AI-complementary perceptive skills using the seed idea?
3. How successfully does this practice support the student in generating sustainable entrepreneurial insights?

In the current stage of implementation, teachers relied primarily on qualitative observation, reflective discussions, and student outputs recorded in digital portfolios to infer the progression of AI- complimentary perceptive skills. The next stage will build upon these practices by transforming the existing portfolio process into a structured, triangulated assessment framework to formalize teacher observations and strengthen reliability of assessment.

In future iterations, this framework will include observations from rubrics, reflective portfolios, tasks for ethical reasoning, and colleague evaluations within this framework. This multimodal assessment method accounts for observable behaviours and inward cognitive–ethical development while strengthening reliability in the assessment of human-centred competencies that accompany AI cognition.

3.4 AI-Native Experimentation

The physics laboratory is designed so that every seed idea can be explored experimentally. This is achieved by correlating each seed idea with specific experiments set up in the laboratory. For example, Melde's string experiment is linked to the following ten seed ideas:

1 Every elastic material can function as an oscillator, capable of producing, transmitting, and receiving waves.
2 Oscillation occurs when a body attempts to return to equilibrium after being displaced by an external force.
3 An undamped oscillator oscillates indefinitely at its natural frequency (the maximum achievable frequency) while maintaining a constant amplitude.
4 A damped oscillator oscillates at frequencies lower than its natural frequency, with the frequency proportional to its damping level.
5 The dissipation of energy in an oscillator depends solely on the damping forces acting on it.
6 A damped oscillator requires energy input from external sources in the form of waves to sustain oscillations; aperiodic forces can disrupt or eliminate its oscillations.
7 A cause-and-effect relationship exists between oscillations and waves: oscillators generate waves, and waves induce oscillations.
8 By adjusting the mode of coupling, the frequency of a wave in a medium can differ from that of the originating oscillator.

9 A rigid surface reflects mechanical waves out of phase, while a non-rigid surface reflects them in phase.
10 Two strategically arranged rigid surfaces can trap a mechanical wave, producing a stationary or standing wave.

Each laboratory experiment is similarly correlated with a set of seed ideas. Experiments are scheduled so that each student's first lab activity involves an experiment related to their unique seed idea. Students are expected to approach this experiment with the entrepreneurial spirit of the original inventor of their seed idea, which they have explored and elaborated on in previous stages. They should apply a holistic perceptive approach, which they have cultivated, to fully grasp the physical phenomenon associated with their seed idea. This understanding will enable them to accurately describe the phenomenon to AI tools for simulation in the subsequent stage.

In essence, students gain an immersive understanding of the phenomenon behind their seed idea through hands-on experimentation. The insights obtained from these experiments serve as a foundation for further learning and provide a benchmark for data they may later obtain from AI tools.

Assessment Criteria:

1. Imaginative ability to correlate the seed idea with the physical quantity measured during the experiment.
2. Skill in reinterpreting experimental results in the context of their seed idea.
3. Capacity to derive a new entrepreneurial insight based on the experiment.

3.5 AI-Native Simulation

The laboratory experiment is immediately followed by a simulation of the seed idea using AI tools. Armed with a deep, immersive understanding of the physical phenomenon behind their seed idea, students are well-positioned to prompt AI tools to generate the corresponding simulation. Students are free to choose any programming language to develop code that simulates the phenomenon they identified during their laboratory experiment. They are supported in selecting the most suitable AI tool, which can generate the best code in their chosen programming language.

The developed code is validated against the data students recorded during their laboratory experiments. The experimental dataset is assumed to reflect the 'fingerprints' of the physical phenomenon associated with their seed idea. Therefore, if the AI-generated code can replicate the experimental dataset, the code is considered valid, as it accurately captures the physical phenomenon.

This approach to code development is designed to prepare students for the workspaces of tomorrow, where they will be tasked not with the routine generation of code, but with the precise understanding of requirements and using AI tools to generate appropriate code accordingly. In other words, the focus shifts from writing code to immersively understanding and describing the phenomenon to AI tools, and then validating the results against their own observations. Students are equipped in both AI-collaborative and AI-complementary ways in relation to tasks involving simulation and automation.

Assessment Criteria:

1. Imaginative Power: Ability to identify the correct correlation between the seed idea and the physical quantities measured in the laboratory.
2. Future-readiness: Ability to use AI tools, such as ChatGPT, to generate appropriate code.
3. Critical Thinking: Ability to interpret the results of the simulation.

3.6 Development of AI-Native Thinking Skills

AI tools are equipped with algorithms that can either replace or complement human cognitive skills. Just as students have distinct learning styles, AI tools are characterized by the learning algorithms they employ. In this context, the intellectual formation of AI natives must be reimagined to include not only thinking skills already mastered by AI tools, but also those that AI has not yet fully replicated and may only approximate without consciousness or ethical intent. Training students exclusively in skills where AI excels may render them vulnerable in future workspaces. Jobs suited to AI are likely to remain within its exclusive domain, potentially limiting human access to these roles. By contrast, training in skills that allow collaboration with AI offers students opportunities to engage in tasks such as managing AI-based systems or working alongside AI on specific tasks.

Therefore, it is essential to distinguish between AI-collaborative thinking skills—those where AI has already achieved mastery and can enhance human efficiency—and AI-complementary thinking skills—those that currently remain rooted in distinctive human dimensions such as awareness, empathy, intentionality, and ethical reasoning in which humans complement AI as part of the thinking process. In shaping the intellect of AI-native students, training in AI-collaborative skills is a necessary condition to keep humans involved in AI-driven processes, while training in AI-complementary skills is a sufficient condition to differentiate human capability. Thus, modern engineering education should cultivate both AI-collaborative and AI-complementary thinking skills.

The cognitive skills in the Revised Bloom's Taxonomy (RBT) [23] are classified as AI-collaborative, as AI tools have already achieved relative proficiency in these skills. They can be categorized into two types: 1) Endo-personal thinking skills: These skills involve retrieving existing knowledge within an individual and include the lower-order thinking skills (LOTs) of RBT, such as remembering, understanding, and applying knowledge. 2) Exo-personal thinking skills: These skills involve engaging with external information or data and include the higher-order thinking skills (HOTs) of RBT, such as analysing (analytical thinking), evaluating (critical thinking), and creating (creative thinking).

As noted by [24], AI natives must develop thinking skills that go beyond those outlined by RBT. AI-complementary thinking skills, which AI tools are unlikely to replicate, fall into two main categories: 1) Interpersonal Thinking Skills: These facilitate the communication of knowledge between individuals. It includes: A) Mutual thinking: Skills for direct communication, such as prompt engineering for AI collaboration, are likely to remain a human core competence, even as AI masters business-like communication. B) Figurative thinking: Skills for transferring meaning across levels—like using figures of speech—are quintessentially human and unlikely to be replicated by AI. C) Imaginative thinking: Skills for designing worldviews, creating visions, or conveying a philosophy of life, critical for thought leadership, remain unique to humans.

2) Transpersonal Thinking Skills: These enable individuals to spontaneously acquire knowledge from unexpected sources and often involve sudden insights. Serendipity, or the reception of knowledge by prepared minds, characterizes these skills. It includes: A) Intuitive thinking: The ability to receive novel ideas, such as Archimedes' and Newton's insights, requires mindfulness and remains a deeply human skill. B) Revelatory thinking: Insights received without full understanding, as seen in Ramanujan's work, rely on openness and meditativeness. C) Mystical thinking: Profound, resonant knowledge about reality often transcends verbalization, remaining out of reach for AI.

To succeed in AI-dominated workplaces, engineering students need intellectual formation that distinguishes them with uniquely human thinking skills, while enabling them to work creatively alongside AI. The emphasis should be on developing AI-complementary thinking skills alongside AI-collaborative ones, ensuring human thinking remains relevant and resilient.

In the proposed curriculum, students gain AI resilience by practicing both AI-collaborative and AI-complementary thinking using AI tools themselves. Although AI may not access unique human thinking dimensions, it can assist students in systematically nurturing these skills. Through reflective work on their seed ideas, students are trained to master both AI-collaborative and AI-complementary skills, equipping them for success in AI-driven environments.

Assessment criteria:

1. How well does the student develop AI-collaborative thinking skills using the seed idea?
2. How well does the student develop AI-complementary thinking skills using the seed idea?
3. How effectively does this exercise contribute to the student's ability to generate impactful entrepreneurial insights?

3.7 Entrepreneurial Sprouting

An intense and personal engagement with their own seed idea (or idea egg)—starting from its harvest, moving through its AI-native elaboration, application in developing AI-native perceptive skills, immersive understanding through AI-native experimentation, AI-native simulation, and finally its application in AI-native thinking skills—aims to facilitate its growth into an entrepreneurial insight within a designated incubation period. In the Entrepreneurial Physics curriculum, three weeks are allotted for the incubation of the seed idea into an entrepreneurial insight, drawing a parallel to the typical hatching period of a mother bird in India's climate. In this stage, the first intrinsic learning motivation, namely the joy of knowledge creation, finds application.

In this process, seed ideas, which cover the full syllabus of the Engineering Physics program, serve dual purposes in the entrepreneurial education of AI natives: 1) Entrepreneurial Personal Transformation: This cultivates an entrepreneurial mindset, value set, skill set, and life set, equipping students to navigate the volatility, uncertainty, complexity, and ambiguity (VUCA) introduced by AI tools in their personal and professional lives. 2) Entrepreneurial Transformation of the World Around: This transformation is fuelled by the entrepreneurial insight each student develops through their intimate engagement with their unique seed idea. Such insights can grow into projects or solutions that sustainably impact their family, community, village, state, nation, or even the broader globe. In short, the entrepreneurial transformation of Engineering Physics (Entrepreneurial Physics) fulfils two core objectives of entrepreneurial education: the transformation of the individual learner and the positive transformation of the world around them, making entrepreneurial education a sustainable and impactful practice.

The entrepreneurial transformation of education, particularly engineering education, reshapes how AI natives are taught, contributing to their personal and community-oriented development into entrepreneurial thinkers. This approach enables education to empower students and all other educational stakeholders, fostering positive societal impact through the entrepreneurial learning outcomes generated by student engagement with curriculum content. Integrating this process into the curriculum also alleviates students' need to engage in unrelated co-curricular, extracurricular, and intramural activities, which might otherwise distract them from their studies. With entrepreneurial education embedded within the curriculum, students are expected to develop intrinsic learning motivation, making their educational journey autonomous, lifelong, venturesome, and flexible—qualities essential for managing the VUCA-driven workspaces of the AI era.

3.8 Entrepreneurial Skill Development

In an age of acceleration, AI natives must master the ability to adapt their skillset on demand. Realizing an entrepreneurial insight often requires new skills, and students are encouraged to freely develop and add skills as part of becoming future-ready in the AI era. In this stage, students are challenged to cultivate the skills necessary to bring their entrepreneurial insights—formed through ongoing engagement with their seed ideas—to life. They are encouraged to identify and develop any skill they deem essential for this purpose. In this stage, the second intrinsic learning motivation, namely the joy of self-actualization, is expected to be amply generated and consumed.

The goal of this stage is to instil in AI natives an understanding of the critical need for flexible learning. The most vital skill in the VUCA (Volatile, Uncertain, Complex, Ambiguous) age is the ability to evolve one's skillset as needed. The entrepreneurial skill development process is powered by intrinsic motivation, driven by the joy of self-actualization.

Students work with AI tools to brainstorm and generate a list of skills required to actualize their entrepreneurial insights, ultimately selecting a skill to pursue. Once chosen, they work to internalize this skill through rigorous training, building it into their "muscle memory", and adding it into their skillset. AI tools may assist in conceptualizing their training program, but students themselves are responsible for the actual skill development..

Assessment Criteria:

1. Skill Originality: Originality in connecting the entrepreneurial insight to the skill identified.
2. Skill Mastery: The level of expertise demonstrated in the developed skill.
3. Skill Utility: Practical usefulness of the developed skill for its intended purpose.

3.9 Synergization of Startup Idea

While students have worked individually with their seed ideas up to this stage, they now transition to working in startup teams. These teams are intentionally designed by instructors to capture the full diversity of the classroom, including variations in gender and academic performance. Students work within these instructor-engineered teams to prepare for generating entrepreneurial outcomes, mirroring the diverse teams they may encounter in future workspaces. Each team member has already demonstrated entrepreneurial efficacy by successfully developing insights from randomly assigned ideas, ensuring that all members are capable of contributing meaningfully.

The startup teams are tasked with creating impactful startup concepts, fuelled by collaborative synergy. The goal is for the startup idea to emerge as a shared brainchild, so all members can claim authentic ownership. This synergistic process is driven by the intrinsic motivation of joyful collaboration.

To keep the process manageable within the classroom environment, the general product focus is set on developing educational toys designed to promote mastery in Science, Technology, Engineering, Arts, and Mathematics (STEAM) for schoolchildren. This focus eliminates the need for extensive market research, allowing students to concentrate on innovation, keeping the program within the time constraints of the semester. Throughout this stage, students are encouraged to leverage AI tools to support various aspects of the development process.

This stage of entrepreneurial curricular process is fostered by two perennial intrinsic learning motivations, namely joy of self-expression and joy of playful synergy.

Assessment Criteria:

1. Innovation Value: How original and creative is the STEAM toy concept?
2. Futuristic Value: How impactful and forward-thinking is the toy?
3. Social Impact: What is the potential effect of the toy on children's learning?

3.10 Simplification of Startup Idea

In this stage, students are required to radically simplify their startup ideas to enhance sustainability. According to the principle of natural selection, only the most simplified, streamlined, and well-adapted ideas survive. Thus, startup teams focus on making their innovations frugal enough to promote long-term sustainability. To achieve this, they employ gamification to simplify their startup ideas, creating a 'game-twin' version of their concept and actively engaging with it as players.

As outlined by [25] in his exploration of gamifying physics themes, gamification can help reveal ways to make an idea more engaging and impactful. Through their interaction with the game version of their startup idea, teams can discover ways to enhance its appeal and simplicity. Additionally, they create a junior version of the game tailored for the age group targeted by their STEAM toys, observing how children engage with this version. These insights are expected to refine and optimize their startup ideas further.

Students are encouraged to consult AI tools to aid in the gamification process. AI can be especially useful in developing age-appropriate features for the game, offering data-driven insights to optimize the game's design for their intended audience.

Criteria of Assessment:

1. Level of Synergy: To what extent does the game concept reflect the collective insights of team members?
2. Level of Fun: How much enjoyment and insight does the game generate?
3. Level of Simplification: How elegant and simple is the game design?

3.11 Pitching of Startup Idea

The pitching of the startup idea takes place before a 'Shark Tank' panel, comprising representatives from nearby industries, makerspaces, and incubation centres. This event is organized by the Innovation and Entrepreneurship Development Cell (IEDC) of Rajagiri School of Engineering & Technology, in collaboration with the institute's Maker Space. These entities are involved to ensure that the entrepreneurial spirit cultivated in the Engineering Physics course permeates the broader engineering curriculum at RSET, fostering a culture of innovation and entrepreneurship across the program. This stage in the entrepreneurial curricular process feed on the intrinsic learning motivation associated with the joy of selfless service.

During the pitch, startup teams aim to attract investment for their startup ideas by presenting a prototype of the STEAM toy they intend to commercialize. The investment offered by the Shark Tank is assumed to reflect the market value of the entrepreneurial outcome produced by the team, highlighting their collective entrepreneurial potential. Consequently, students' grades at this stage are aligned with the percentile of investment their startup receives from the panel.

To prepare for this presentation, students are encouraged to leverage AI tools to develop the most effective pitch. They work with AI both collaboratively and in complementary ways to enhance their pitch.

Criteria of Assessment:

1. Innovation Value: How innovative is the startup concept, as assessed by the Shark Tank panel.
2. Futuristic Value: How forward-thinking is the solution developed by the team, as judged by the panel.
3. Social Impact: The potential societal impact of the startup idea, as stated by the team and evaluated by the panel.

3.12 Entrepreneurial Assessment

An assessment framework for entrepreneurial education tailored to AI natives is presented in [26]. Throughout the course, students preserve each activity in the curriculum within digital portfolios. They have the opportunity to refine and improve their work at each stage until the course concludes, reinforcing that meaningful engagement with the curriculum is valued over mere task completion. At the end of the course, the digital portfolios are assessed by AI tools designed specifically for this purpose, using assessment criteria for each stage previously communicated to the students. This approach ensures an objective evaluation of student performance. The final scores for each student are aligned with their demonstrated ability to transform both themselves and the world around them in the spirit of entrepreneurship, affirming the AI resilience they have developed during the course.

The curricular process described in this paper treats education as a process through which students of engineering develop their self-actualization and self-efficacy as would-be engineers. As they subject themselves to diverse learning experiences in different courses, whose curricular processes are patterned in the manner described in this paper, they progress along the path of becoming engineers in the Age of Intelligence, governed by AI technology.

4 Results

This AI-resilient curricular process has been evolving over nearly three years at Rajagiri School of Engineering & Technology, India. Approximately 3,000 engineering students have already been trained using this pedagogy, which is grounded in students' intrinsic learning motivations and directed toward entrepreneurial learning outcomes. Psychometric tests, developed based on student feedback, indicate significant improvements in learning motivation as a result of participating in the Engineering Physics course designed according to this curricular process. The process is continually assessed and optimized to enhance its effectiveness.

The primary goal of the newly defined curricular process is to enhance the AI resilience of engineering students by equipping them to collaborate with AI tools while simultaneously developing uniquely human skills that AI does not authentically replicate. The curriculum encourages students to engage with AI tools as thinking assistants and creativity companions. Students are progressively trained to use AI tools effectively, mastering the art of prompt engineering as they advance through each stage of the program. Concurrently, they are required to cultivate perceptive and critical thinking skills that remain beyond the capabilities of AI tools.

A key aspect of the process involves students developing useful or entrepreneurial outcomes. Student startup teams are given the opportunity to pitch their ideas to prospective investors in a Shark Tank-style event, serving as the program's culminating activity. Representatives from industries who participated in the initial Shark Tank event expressed significant interest in future participation, underscoring the entrepreneurial value of the learning outcomes demonstrated by the students.

Student learning experiences are documented in digital portfolios, which are subsequently subjected to AI-based assessments to complete the curricular process. These assessments indicate that the students effectively achieve the targeted learning outcomes by following the curricular methodologies described in this paper.

5 Limitations and Outlook

This study was conducted on a course titled Engineering Physics, which is considered a foundational subject for all engineering students. Although the course is designed to provide a scientific foundation for various engineering disciplines, it is not classified as a core engineering discipline. Therefore, it is essential to extend this research to core engineering disciplines to evaluate its effectiveness in those contexts.

A typical engineering curriculum comprises various types of courses designed to achieve globally defined Program Outcomes (POs) for engineering education. Consequently, this study must be expanded to include these diverse courses within the engineering curriculum before proposing this approach as a foolproof method for fostering AI resilience in modern engineering students.

Future prospects for this research include expanding the proposed pedagogical and curricular processes into different stages of education, starting with the formative years in preschool education. This approach could also be tested in school-level and higher education. Furthermore, the same methodology could be applied to ongoing educational interventions beyond university education.

In essence, this entrepreneurial mode of education—centred on intrinsic learning motivations and aimed at generating entrepreneurial learning outcomes—has the potential to transform the pedagogy, andragogy, and heutagogy of modern learners in the AI-driven Age of Intelligence. Systematic research must build upon this foundational work to bring about a paradigm shift in educational practices for the AI era.

6 Conclusion

Building AI resilience is essential for the next generation of students, often referred to as AI natives, to safeguard their uniquely human competencies in perception and critical thinking. As AI tools increasingly master these capabilities, they pose an existential challenge to AI natives in the workplace. To address this, modern engineering students must cultivate both AI-collaborative and AI-complementary perceptive and cognitive skills to maintain a competitive edge in their careers.

This paper presents insights from an experimental implementation of a specialized pedagogy for AI natives, integrated into the Engineering Physics curriculum for first-year engineering students. The findings demonstrate that fostering AI resilience in engineering students is achievable through the effective use of AI tools, coupled with adequate mastery of prompt-engineering techniques.

Instructors utilized AI tools to help students generate and refine initial "seed ideas," enabling detailed, original, interdisciplinary, and historically grounded elaborations of each assigned idea. Through this structured curricular process, students developed both AI-collaborative and AI-complementary perceptive skills, achieving a comprehensive understanding of their topics. Immersive experimentation with their seed ideas further allowed students to uncover the core scientific phenomena underlying these concepts. AI tools played a key role in simulating these phenomena, simultaneously enhancing students' coding skills and fostering their understanding of the principles of physics involved.

Additionally, students honed their AI-collaborative and AI-complementary thinking skills by using their seed ideas as cognitive content. AI tools were employed to support this process, enabling students to build AI resilience. These tools also acted as creativity boosters, facilitating brainstorming sessions and optimizing entrepreneurial insights derived from the seed ideas.

Students used AI to identify the skill adaptations necessary to bring their entrepreneurial ideas to fruition and received guidance on mastering these skills. Within startup teams, AI tools supported the synergistic development of innovative ideas, promoting engagement and collaboration. AI was also instrumental in simplifying these ideas for sustainability and in crafting compelling pitches to attract potential investors.

Throughout this process, students documented their progress in portfolios, which included records and reflections on each step of this entrepreneurial pedagogy. These portfolios were assessed using AI tools to ensure objective evaluation and feedback.

The seamless integration of AI tools into the Engineering Physics curriculum highlights how these resources can be effectively introduced to engineering students to cultivate the AI resilience necessary for navigating the volatile, uncertain, complex, and ambiguous (VUCA) environments of the AI-driven age. This study aims to inspire further research and innovation in transforming educational practices across all levels and forms, ensuring a robust, AI-resilient future.

References

1. Chen, L., Chen, P., Lin, Z.: Artificial intelligence in education: A review. IEEE Access **8**, 75264–75278 (2020)
2. Woerner, J.H.R., Turtova, A.P., Lang, A.S.I.D.: Transformative Potentials and Ethical Considerations of AI Tools in Higher Education: Case Studies and Reflections. In: Proc. SoutheastCon, pp. 510–515 (2024)
3. Panthalookaran, V.: A Revised taxonomy of educational objectives for AI-Natives: Reimagining thinking skills in the age of AI. Higher Education for the Future, p. 23476311241285100 (2024)
4. Walter, Y.: Embracing the future of Artificial Intelligence in the classroom: the relevance of AI literacy, prompt engineering, and critical thinking in modern education. Int. J. Educ. Technol. High. Educ. **21**(1), 15 (2024)
5. Bravo, F.A., Cruz-Bohorquez, J.M.: Engineering Education in the Age of AI: Analysis of the Impact of Chatbots on Learning in Engineering. Education Sciences **14**(5), 484 (2024)
6. Khan, T.A., Alam, M., Rizvi, S.A., Shahid, Z., Mazliham, M.S.: Introducing AI applications in engineering education (PBL): An implementation of power generation at minimum wind velocity and turbine faults classification using AI. Comput. Appl. Eng. Educ. **32**(1), e22691 (2024)
7. Onatayo, D., Onososen, A., Oyediran, A.O., Oyediran, H., Arowoiya, V., Onatayo, E.: Generative AI Applications in Architecture, Engineering, and Construction: Trends, Implications for Practice, Education & Imperatives for Upskilling—A Review. Architecture **4**(4), 877–902 (2024)
8. Xie, M., Meng, F., Zou, J., Feng, W., Ma, S.: Application of artificial intelligence in civil engineering education. J. Phys.: Conf. Ser. **1575**(1), 012225 (2020)
9. Ambikairajah, E., Sirojan, T., Thiruvaran, T., Sethu, V.: ChatGPT in the Classroom: A Shift in Engineering Design Education. In: Proc. 2024 IEEE Global Engineering Education Conf. (EDUCON), pp. 1–5 (2024)
10. Mokhtarpour, A., de Freitas, C.C.S.: Engaging First-Year Engineering Students: A Technology-Based Approach Using Story-Based Learning and AI-Generated Content. in Proc. 2023 IEEE Global Humanitarian Technology Conf. (GHTC), pp. 358–365 (2023)
11. Cady, S.H., Willing, J.G., Cady, D.A.: The AI Imperative: On Becoming Quintessentially Human. J. Appl. Behav. Sci. **60**(1), 1–20 (2024)
12. Glory, J.S., Nalina, E., Santhosh, Y., Prasad, D.: Enhancing holistic education through AI-integrated learning environments: A comprehensive framework. In: Proc. 2023 Int. Conf.

Res. Methodologies Knowl. Manage. Artif. Intell. Telecommun. Eng. (RMKMATE), pp. 1–5 (2023)
13. Jeon, I.-S., Yun, S.-Y., Shin, S.: Examining AI and digital ethics issues for youth and educational approaches: Focusing on IoT and generative AI. The Journal of Learner-Centered Curriculum and Instruction **24**(10), 537–550 (2024)
14. Khan, M.A., Kurbonova, O., Abdullaev, D., Radie, A.H., Basim, N.: Is AI-assisted assessment liable to evaluate young learners? Parents support, teacher support, immunity, and resilience are in focus in testing vocabulary learning. Language Testing in Asia **14**(1), 48 (2024)
15. Lake, B.M., Ullman, T.D., Tenenbaum, J.B., Gershman, S.J.: Building machines that learn and think like people. Behavioral and Brain Sciences **40**, e253 (2017)
16. Piaget, J.: The Origins of Intelligence in Children. International University (1952)
17. Panthalookaran, V.: A Blueprint for an AI-resilient education. In: EDULEARN24 Proceedings (2024)
18. Panthalookaran, V.: A blueprint for pedagogy of digital natives in the AI Age. In: EDULEARN24 Proceedings (2024)
19. Panthalookaran, V.: Pedagogy of the digital natives: An introduction to entrepreneurial education. Pallikkutam Publications (2023)
20. Panthalookaran, V.: Education in a VUCA-driven world: Salient features of an entrepreneurial pedagogy. Higher Education for the Future **9**(2), 234–249 (2022)
21. Panthalookaran, V.: On integrating entrepreneurial pedagogy into physics education for engineers. In: ICERI2023 Proceedings (2023)
22. Panthalookaran, V.; Towards a universal curriculum for entrepreneurial education based on intrinsic learning motivations. In: 15th Biennial Conference on Entrepreneurship, Entrepreneurship Development Institute of India (EDII) (2023)
23. Anderson, L.W., Krathwohl, D.R.: A taxonomy for learning, teaching, and assessing: A revision of Bloom's Taxonomy of educational objectives: Complete Edition. Addison Wesley Longman, Inc. (2001)
24. Panthalookaran, V.: Beyond Bloom's taxonomy: Emergence of entrepreneurial education. Higher Education for the Future **9**(1), 45–61 (2022)
25. Panthalookaran, V.: Gamification of physics themes to nurture engineering professional and life skills. 2018 IEEE Global Engineering Education Conference (EDUCON) (2018)
26. Panthalookaran, V.: An assessment framework for developing an entrepreneurial mindset in next-generation learners. ICERI2023 Proceedings (2023)
27. Panthalookaran, V., Sotiriadou, C., Zinn, B.: Psychometric test for measuring the growth of intrinsic learning motivations in students. Submitted to Higher Education for the Future (2024)

A Comparative Study of AI-Driven and Human Tutoring Models for Coding Instruction in Initial Teacher Education

Stavros A. Nikou[1(✉)], Maria Perifanou[2], and Anastasios A. Economides[2]

[1] University of Strathclyde, Glasgow, UK
stavros.nikou@strath.ac.uk
[2] University of Macedonia, Thessaloniki, Greece

Abstract. The study investigates the impact of human tutor support versus Generative AI (GenAI) support on preservice teachers' learning of basic programming concepts, explicitly situating GenAI in the context of preservice teacher education rather than conventional computer science cohorts. The study follows a quasi-experimental research design. 70 preservice teachers from a teacher education program at a UK university were divided into two groups: a control group that received traditional in-class support from a human tutor, and an experimental group that received equivalent support from a GenAI system during structured sessions. Quantitative data from the online survey was analyzed using descriptive and inferential statistics. Cognitive engagement, enjoyment and expected performance were found higher for the control group while satisfaction was found higher for the experimental group. No significant differences were found for self-efficacy, competence support, behavioral engagement, emotional engagement, and intention to use. The findings suggest that human tutors foster deeper cognitive engagement and enjoyment, while GenAI enhances satisfaction and autonomy. This complementarity suggests that integrating both forms of support can improve programming education in teacher education programmes by leveraging the unique affordances of each.

Keywords: programming · Generative AI · human tutoring · student teachers

1 Introduction

As digital literacy and computational thinking have become essential elements of education, it is increasingly important for preservice teachers to master basic programming skills [1, 2]. However, many preservice teachers come from non-technical backgrounds, and often they face challenges in learning programming [3, 4]. Human tutors' support is the mainstream method for overcoming these challenges. With the recent advancements in Generative Artificial Intelligence (GenAI), there is a growing interest in employing AI-powered tools to support or even replace human tutoring in educational settings. However, there is little empirical evidence on preservice teachers' perceptions about the role of GenAI in supporting programming education [5]. The current study aims to

M. E. Auer et al. (Eds.): IMCL 2025, LNNS 1954, pp. 59–68, 2026.
https://doi.org/10.1007/978-3-032-24998-2_6

address this gap by exploring the perceptions of preservice teachers who learn basic programming through either human tutor support or GenAI support. This approach is novel in situating Generative AI support within preservice teacher training rather than typical computer science cohorts. Unlike prior studies focused on undergraduates, it directly compares human tutoring and GenAI tutoring in a controlled, quasi-experimental design with preservice teachers learning Scratch. By analyzing perceptions across cognitive, emotional, and motivational dimensions, it advances understanding of how GenAI complements human tutors in preparing non-technical teacher trainees for programming pedagogy. The findings can inform the design of teacher training programmes on teaching basic programming and contribute to the discussion about integrating GenAI in programming education.

2 AI Tools in Programming Education: A Literature Review

The integration of artificial intelligence (AI) in programming education represents a significant paradigm shift in how students learn computational skills. Research demonstrates that AI tools can substantially enhance the educational experience for novice programmers, though this integration presents both opportunities and challenges that require careful consideration.

Previous studies found that students exploited AI across diverse programming tasks, including code generation, debugging, conceptual inquiries, and test case creation [6]. Students asked AI tools for help with debugging and conceptual questions more often than requesting complete code solutions [7], suggesting a constructive approach to AI integration where students seek understanding rather than merely copying answers. Also, students utilized AI tools to get started with programming assignments, generate text and code, assist with debugging, and serve as comprehensive search tools [8]. Furthermore, AI helped students acquire background knowledge, learn syntax and programming concepts, and receive suggestions for suitable algorithms [9].

AI integration in programming education enhanced student engagement and learning experience. Students reported more positive learning experiences and deeper engagement with learning content [10]. AI proved particularly valuable for reluctant learners, helping them explore programming concepts they previously considered beyond their skill level [11]. Additionally, AI tools significantly improved students' self-regulated learning abilities, including intrinsic goal orientation, task value, and metacognitive self-regulation [12]. Effective collaboration between students and AI can enhance both individual learning and human-to-human communication skills [13].

Students who utilized AI in programming education consistently demonstrated superior learning outcomes across multiple dimensions. Previous studies found that AI users achieved higher computational thinking skills, enhanced programming self-efficacy, and increased motivation for learning [14]. Most significantly, students using AI tools achieved statistically significantly higher coding scores while showing comparable improvements in understanding fundamental programming concepts [12]. AI tools exceled particularly in debugging and code comprehension tasks. Students using AI could fix programming bugs more quickly and effectively than those working without assistance [10]. AI-generated code explanations were consistently rated as significantly easier to understand and more accurate than student-created explanations [15]. Furthermore, AI helped students understand fundamental programming concepts and identify common syntax and semantic errors [16].

Despite these benefits, significant challenges accompany AI integration in programming education. A critical concern was the negative correlation between students' performance and the reported frequency of AI usage [17], suggesting that excessive dependence may hinder learning progression. Students and educators expressed concerns about over-reliance on AI tools and potential academic integrity issues [18]. Technical limitations present additional challenges. Students reported that AI tools sometimes provide incomplete or incorrect answers, causing professional anxiety about the reliability of AI-generated solutions [19]. The AI integration may also accidentally promote academic laziness, as students become accustomed to seeking immediate solutions rather than engaging in problem-solving processes [Yilmaz & Yilmaz, 2023a].

Concluding, AI integration in programming education offers substantial benefits in enhancing learning outcomes, debugging capabilities, and student engagement. However, much of this research on programing education has been deployed in computer science and other undergraduate or graduate courses while little research exists in programming education using GenAI in preservice teacher courses [20]. Programming education for preservice teachers is distinct because it mainly aims to build confidence in student teachers to teach programming instead of actually do programming. Moreover, since many preservice teachers come from non-technical backgrounds, they may not be motivated enough to engage in programming [21]. Therefore, exploring student teachers' perceptions and engagement with different teaching strategies for programming, e.g., human-led versus GenAI-supported instruction, is essential for understanding how best to support their learning and future classroom practice.

3 Methodology

3.1 Participants

A total of 70 preservice teachers from a teacher education program at a Scottish University participated in the study. Participants had very limited or no prior experience in programming or computational thinking. The study employed a quantitative quasi-experimental design. Participants were randomly assigned to one of two groups: 38 in the control group (tutor-supported) and 32 in the experimental group (GenAI-supported). Ethical approval was secured prior to the study from the researchers' University Ethics Committee, and the study has been conducted during the fall 2024. Participation was on a voluntary basis and informed consent was obtained from all participants.

3.2 Procedures

The intervention lasted for four weeks, with one 2-h session conducted each week, resulting in a total of eight hours of instruction. All participants used Scratch, a visual block-based programming language, widely accepted as an appropriate environment to learn basic programming concepts. The learning objectives were student teachers to understand the basics of sequential and event-driven programming, learn how to use variables, conditional statements and simple loops. The instructional content and learning activities were consistent across both groups to ensure comparability. To address the above objectives, a series of tasks were introduced. Students created a sprite that moves across the screen when a key is pressed, made the sprite repeat dance figures using loops, created a quiz that asks a question and responds based on the answer. Also, students were asked in the final session to combine all the above by creating a simple game e.g., a sprite that moves through a maze or a pong game. During the above learning tasks, the control group was supported by a qualified human tutor during each session. The tutor provided guidance, answered questions, and facilitated problem-solving in a traditional face-to-face format allowing students to use online (non-GenAI) tutorials and manuals. The experimental group received support via a GenAI interface (Microsoft Copilot) that responded to natural language queries, provided hints, explained concepts, and offered code suggestions in real time. To ensure comparability and internal validity all instructional and support interactions occurred in a controlled setting, with participants receiving guidance only during the weekly 2-h sessions under research supervision. Students in the experimental group accessed GenAI only under in-class supervision, while the control group worked with a tutor under the same conditions without using GenAI for learning programming outside the class.

3.3 Instruments

At the conclusion of the intervention, participants completed a post-intervention online survey designed to assess variables related to their experiences. The survey was adapted from previously validated instruments in educational technology and computing education research to ensure reliability and construct validity. The survey comprised Likert-scale items evaluating the following constructs.

Programming self-efficacy is based on the notion of individual's belief in their ability to succeed in specific situations or accomplish a task [22] assesses the learner's confidence in their ability to complete coding tasks. Example item was "I can complete a new programming task that is assigned to me if I keep trying". Competence support refers to the resources and assistance provided to help users perform a task and, in our context, how GenAI or human tutors' support enhances learning programming. Example item was "Using GenAI tutor/human tutor makes me feel like I am good at programming". Behavioural engagement refers to learners' active observable participation in learning activities [23, 24]. Example item was "I put effort into learning programming concepts when working with the GenAI/human tutor". Cognitive engagement refers primarily to the strategies used by learners to complete a learning activity [23, 24]. Example item was "I go through my programming work with the GenAI/human tutor and make sure that my solutions are correct". Emotional engagement refers to the experience of positive or negative emotions during a learning activity. Example item was "I feel good when I learn programming with GenAI/human tutor support".

Satisfaction and enjoyment are the levels of learners' overall satisfaction and enjoyment with the support modality. For satisfaction, we adopted items from [25] and for enjoyment we adopted items from [26]. Example items are "I feel satisfied when I complete my programming assignment" and "Using GenAI/human tutor support in learning programming is enjoyable". Intention to use was adopted from [27], e.g., "I intent to deploy GenAI/human tutor to help me in learning programming". Expected performance (perceived learning), refers to the degree to which a person believes that using the system can help attain goals in terms of job performance [17] for example, "I believe my programming skills have improved as a result of this course".

4 Data Analysis and Results

Quantitative data from the online survey was analyzed using descriptive and inferential statistics using IBM SPSS v.28. Internal consistency of each survey scale was measured using Cronbach's alpha. Cronbach's alpha for the 3 programming self-efficacy items was 0.713, for the 3 competence support items 0.736, for the 3 behavioral engagement items 0.864, for the 4 cognitive engagement items 0.742, for the 3 emotional engagement items 0.711, for the 4 satisfaction items 0.867, for the 3 enjoyment items 0.914, for the expected performance 3 items 0.719 and for the 3 items of the intention to use 0.911. Table 1 presents the descriptive statistics for variables under investigation.

Table 1. Mean rank and median for control and experimental groups.

	Groups	N	Mean Rank	Median
Programming Self-Efficacy	1.00	38	35.43	3.67
	2.00	32	35.58	3.67
Competence Support	1.00	38	31.99	3.33
	2.00	32	39.67	3.66
Behavioural Engagement	1.00	38	35.88	3.50
	2.00	32	35.05	3.33
Cognitive Engagement	1.00	38	30.87	3.50
	2.00	32	41.00	3.87
Emotional Engagement	1.00	38	35.74	3.85
	2.00	32	35.22	3.69
Satisfaction	1.00	38	40.30	4.00
	2.00	32	29.80	3.25
Enjoyment	1.00	38	31.03	3.67
	2.00	32	40.81	4.00
Expected Performance	1.00	38	30.05	3.50
	2.00	32	41.97	4.00
Intention to Use	1.00	38	37.29	4.00
	2.00	32	33.38	3.83

Grouping variable: 1 for experimental group (GenAI), 2 for control group (human tutor)

The median value responses across the above variables were around 3.5 on a 5-point Likert scale, suggesting that the intervention has a rather mild positive impact on attitudes. Enjoyment, satisfaction and intention to use seem to be higher for the intervention (compared to the other values) indicating that support can positively influence learners' emotional engagement and future intention to learn programming.

Since our sample size is relatively small and our ordinal data does not follow a normal distribution (non-parametric), we have used the Mann-Whitney tests to identify significant differences between the two groups across the nine outcome variables.

Mann-Whitney U tests revealed no significant differences for self-efficacy ($U = 605.50$, $p = .976$), competence support ($U = 474.50$, $p = .113$), behavioral engagement $U = 593.50$, $p = .863$), emotional engagement ($U = 599.00$, $p = .915$) and intention to use ($U = 540.00$, $p = .419$).

Significant differences were found for cognitive engagement ($U = 432.00$, $p = .037$), satisfaction ($U = 425.50$, $p = .030$), enjoyment ($U = 438.00$, $p = .043$), and expected performance (perceived learning) ($U = 401.00$, $p = .014$).

Cognitive engagement, enjoyment and expected performance were found higher for the control group with the human tutor support. Satisfaction was found higher for the experimental group with Generative AI support.

Tables 2 and 3 present the inferential statistics for the variables under investigation.

Table 2. Mann-Whitney U test Results for the control and experimental groups

	Programming Self-Efficacy	Behavioral Engagement	Cognitive Engagement	Emotional Engagement
Mann-Whitney U	605.50	593.50	432.00	599.00
Z	-.030	-.172	-2.086	-.106
Asymp. Sig.(2-tailed)	.976	.863	**.037**	.915

Table 3. Mann-Whitney U test Results for the control and experimental groups

	Competence Support	Satisfaction	Enjoyment	Expected Performance	Intention to Use
Mann-Whitney U	474.50	425.50	438.00	401.00	540.00
Z	-1.586	-2.164	-2.023	-2.460	-.807
Asymp. Sig.(2-tailed)	.113	**.030**	**.043**	**.014**	.419

Figure 1 shows the boxplots for the control and experimental groups.

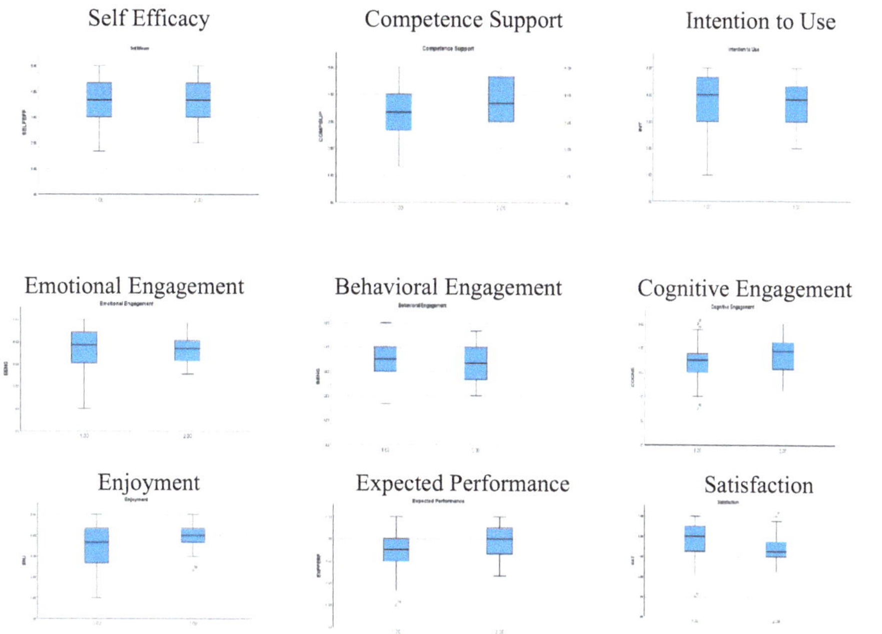

Fig. 1. Boxplots for the variables under investigation, grouping variable: 1 for experimental group (GenAI), 2 for control group (human tutor)

5 Discussions–Conclusion

This study investigated the impact of different support strategies–human tutor support versus Generative AI (GenAI) support-on preservice teachers learning basic programming concepts using Scratch. While GenAI seems promising in enhancing computational thinking and problem-solving abilities [20], there is a lack of research on student teachers' perceptions and engagement in programming education using GenAI support. Unlike most previous studies that examined GenAI use with undergraduate or computer science students, this paper focuses on preservice teachers, a group with limited technical backgrounds and distinct learning goals. It also uses a quasi-experimental control group design, directly comparing human tutor and GenAI support, offering more robust insights into programming education within teacher training contexts.

The research design was a quasi-experimental study involving two groups: a control group receiving traditional tutor support and an experimental group using GenAI-based support. Most of the existing studies on the use of GenAI in programming education do not use a control group [5]; therefore, the incorporation of a control group in our study enhances the robustness of its conclusions. Descriptive and inferential statistical analysis have shown significant differences between the control and experimental groups for cognitive engagement, satisfaction, enjoyment and expected performance (perceived learning). No significant differences were found for self-efficacy, competence support, behavioral engagement, emotional engagement and intention to use. Student teachers in the control group led by human tutors self-reported higher levels of cognitive engagement, enjoyment, and expected performance. Interaction with human tutors seems to require or promote more cognitive engagement compared to interaction with AI. This is probably due to the level of social presence that is conveyed by human tutors. This is in agreement with the social presence theory that suggests that interactive communication in learning environments fosters higher cognitive engagement [28]. The higher level of enjoyment in the human-led group may be attributed to the empathy demonstrated by the human tutors which is difficult to replicate in the GenAI environment. Also, the reported higher level of expected performance in the control group may be due to the personalized, structured and clearer guidance that student teachers received from human tutors. Even though self-efficacy is the same between the groups, tutors in the control group may have better explained to students what was expected from them and how to achieve it and therefore students had a better understanding of the expected performance. In contrast, the higher satisfaction reported by the GenAI-supported group may be due to the novelty of Gen-AI support and the sense of autonomy and control associated with the use of GenAI [29]. Overall, the results suggest that human tutors may better support deeper engagement and confidence, while GenAI may enhance students perceived autonomy, leading to greater overall satisfaction.

Our study findings are in agreement with some of the previous studies regarding the autonomy support associated with the use of GenAI [30] or student satisfaction [31]. However, other studies have shown that students using AI reported deeper engagement with learning content [10] or have achieved higher computational thinking skills, programming self-efficacy, and motivation for learning [12, 14]. These differences with the findings of our study may be due to variations in the type of GenAI used, the age and educational level of the students and of course to the way that GenAI is being used in particular educational contexts.

In conclusion, our findings suggest that human tutors foster deeper cognitive engagement and enjoyment, while GenAI enhances satisfaction and autonomy. This complementarity suggests that integrating both forms of support can improve programming education in teacher education programmes by leveraging the unique affordances of each. Moreover, these dynamics extend beyond coding to other domains where novices face complex learning challenges. In broader educational and professional training contexts, blending human and AI support may optimize engagement, confidence, and learner satisfaction across diverse disciplines.

Exploring the impact of tutor support versus GenAI support on preservice teachers learning programming is an area that needs further exploration with more longitudinal studies considering in more detail how tutors and students interact with and make use of AI tools.

References

1. Lindberg, R.S.N., Laine, T.H., Haaranen, L.: Gamifying programming education in K-12: A review of programming curricula in seven countries and programming games. Br. J. Educ. Technol. **50**(4), 1979–1995 (2019). https://doi.org/10.1111/bjet.12685
2. Tran, Y : Computational thinking equity in elementary classrooms: what third-grade students know and can do. J. Educ. Comput. Res. **57**(1), 3–31 (2019). https://doi.org/10.1177/0735633117743918

3. Durak, H.Y.: Modeling different variables in learning basic concepts of programming in flipped classrooms. J. Educ. Comput. Res. **58**(1), 160–199 (2020)
4. Statter, D., Armoni, M.: Teaching abstraction in computer science to 7th grade students. ACM Trans. Comput. Educ. **20**(1), 8–837 (2020). https://doi.org/10.1145/3372143
5. Cubillos, C., Mellado, R., Cabrera-Paniagua, D., Urra, E.: Generative artificial intelligence in computer programming: does it enhance learning, motivation, and the learning environment? IEEE Access **13**, 40438–40455 (2025). https://doi.org/10.1109/ACCESS.2025.3532883
6. Arora, U., et al.: Analyzing LLM usage in an advanced computing class in India. In Proceedings of the 27th Australasian Computing Education Conference (pp. 154–163) (2025). https://doi.org/10.1145/3716640.3716657
7. Ghimire, A., Edwards, J.: Coding with AI: how are tools like chatgpt being used by students in foundational programming courses. In: International Conference on Artificial Intelligence in Education (pp. 259–267) Cham, Springer Nature Switzerland (2024). https://doi.org/10.1007/978-3-031-64299-9_20
8. Scholl, A., Kiesler, N.: How novice programmers use and experience ChatGPT when solving programming exercises in an introductory course. Proceedings of the 2024 IEEE ASEE Frontiers in Education Conference, Washington, D.C. (2024)
9. Haindl, P., Weinberger, G.: Students' experiences of using ChatGPT in an undergraduate programming course. IEEE Access **12**, 43519–43529 (2024). https://doi.org/10.1109/ACCESS.2024.3380909
10. Koong Lin, H.-C., Tseng, C.-H., Chen, N.-S.: Enhancing programming education: the impact of AI-based pedagogical agents on student self-efficacy, engagement, and learning outcomes. Educ. Technol. Soc. **28**(2), 279–294 (2025). https://doi.org/10.30191/ETS.202504_28(2).TP02
11. Petrovska, O., Clift, L., Moller, F., Pearsall, R.: Incorporating generative AI into software development education. In: Proceedings of the 8th Conference on Computing Education Practice (pp. 37–40) (2024)
12. Huang, A.Y., Lin, C.Y., Su, S.Y., Yang, S.J.: The impact of GenAI-enabled coding hints on students' programming performance and cognitive load in an SRL-based Python course. Br. J. Edu. Technol. (2025). https://doi.org/10.1111/bjet.13589
13. Yan, W., Nakajima, T., Sawada, R.: Benefits and challenges of collaboration between students and conversational generative artificial intelligence in programming learning: an empirical case study. Educ. Sci. **14**(4), 433 (2024). https://doi.org/10.3390/educsci14040433
14. Yilmaz, R., Yilmaz, F.G.K.: The effect of generative artificial intelligence (AI)-based tool use on students' computational thinking skills, programming self-efficacy and motivation. Comput. Educ. Artif. Intell. **4**, 100147 (2023). https://doi.org/10.1016/j.caeai.2023.100147
15. Leinonen, J., et al.: Comparing code explanations created by students and large language models. In: Proceedings of the 2023 Conference on Innovation and Technology in Computer Science Education V. 1 (pp. 124–130) (2023). https://doi.org/10.1145/3587102.3588785
16. Malik, S.I., Ashfque, M.W., Tawafak, R.M., Al-Farsi, G., Ahmad Usmani, N., Hamza Khudayer, B.: A chatbot to facilitate student learning in a programming 1 course: a gendered analysis. Int. J. Virtual Pers. Learn. Environ. **12**(1), 1–20 (2022). https://doi.org/10.4018/IJVPLE.310007
17. Lepp, M., Kaimre, J.: Does generative AI help in learning programming: students' perceptions, reported use and relation to performance. Comput. Hum. Behav. Rep. **18**, 100642 (2025). https://doi.org/10.1016/j.chbr.2025.100642
18. Zastudil, C., Rogalska, M., Kapp, C., Vaughn, J., MacNeil, S.: Generative AI in computing education: Perspectives of students and instructors. In: 2023 IEEE Frontiers in Education Conference (FIE) (pp. 1–9) (2023). https://doi.org/10.1109/FIE58773.2023.10343467

19. Yilmaz, R., Yilmaz, F.G.K.: Augmented intelligence in programming learning: examining student views on the use of ChatGPT for programming learning. Comput. Hum. Behav.: Artif. Hum. **1**(2), 100005 (2023). https://doi.org/10.1016/j.chbah.2023.100005
20. Matobobo, C., Ncube, P.D.N., Ngesimani, N.L., Dzvapatsva, G.P., Chinhamo, E.: Enhancing computational thinking and problem solving in programming education through generative AI: a scoped review. 2025 IEEE Global Engineering Education Conference (EDUCON) (2025)
21. Stupurienė, G., Lucas, M., Bem-Haja, P.: Teachers' perceptions of the barriers and drivers for the integration of Informatics in primary education. Comput. Educ. **208**, 104939 (2024). https://doi.org/10.1016/j.compedu.2023.104939
22. Bandura, A.: Self-efficacy: toward a unifying theory of behavioral change. Psychol. Rev. **84**(2), 191–215 (1977). https://doi.org/10.1037/0033-295X.84.2.191
23. Fredricks, J.A., Blumenfeld, P., Friedel, J., Paris, A.: School Engagement. In: Moore, K.A., Lippman, L.H. (eds.) What Do Children Need to Flourish?. The Search Institute Series on Developmentally Attentive Community and Society, vol 3. Springer, Boston, MA (2005). https://doi.org/10.1007/0-387-23823-9_19
24. Wang, M.-T., Fredricks, J.A., Ye, F., Hofkens, T.L., Linn, J.S.: The math and science engagement scales: scale development, validation, and psychometric properties. Learn. Instr. **43**, 16–26 (2016)
25. Lin, C.S., Wu, S., Tsai, R.J.: Integrated perceived playfulness into expectation–confirmation model for web portal context. Inf. Manage. **42**(5), 683–693 (2005)
26. Deci, E.L., Ryan, R.M.: Handbook of self-determination research. University of Rochester Press (2002)
27. Venkatesh, V., Morris, M.G., Davis, G.B., Davis, F.D.: User acceptance of information technology: toward a unified view. MIS Q. **27**(3), 425–478 (2003)
28. Dunmoye, I.D., Rukangu, A., May, D., Das, R.P.: An exploratory study of social presence and cognitive engagement association in a collaborative virtual reality learning environment. Comput. Educ.: X Reality **4**, 100054 (2024). https://doi.org/10.1016/j.cexr.2024.100054
29. Xie, Z., Wu, X., Xie, Y.: Can interaction with generative artificial intelligence enhance learning autonomy? A longitudinal study from comparative perspectives of virtual companionship and knowledge acquisition preferences. J. Comput. Assist. Learn. **40**(5), 2369–2384 (2024). https://doi.org/10.1111/jcal.13032
30. Annamalai, N., Bervell, B., Mireku, D.O., Andoh, R.P.K: Artificial intelligence in higher education: modelling students' motivation for continuous use of ChatGPT based on a modified self-determination theory. Comput. Educ.: Artif. Intell. **8**, 100346 (2025). https://doi.org/10.1016/j.caeai.2024.100346
31. Liu, C.-C., Liao, M.-G., Chang, C.-H., Lin, H.-M.: An analysis of children' interaction with an AI chatbot and its impact on their interest in reading. Comput. Educ. **189**, 104576 (2022). https://doi.org/10.1016/j.compedu.2022.104576

On Cultivating AI-Native Perceptual Skills in Students

Varghese Panthalookaran(✉)

Department of Basic Sciences and Humanities, Rajagiri School of Engineering & Technology, Kochi 682039, India
varghese@rajagiritech.ac.in

Abstract. As artificial intelligence (AI) becomes increasingly integrated into everyday life and workspaces, it is essential for AI-natives, individuals who are growing up in the AI age, to develop the appropriate perceptual skills necessary for career success. This paper identifies a holistic spectrum of perceptual skills required by AI-natives, delineating the levels of mastery needed based on whether these skills are supported by AI tools or performed independently. In other words, it develops a conceptual framework for distinguishing human perceptual skills that fall within the core competencies of AI tools from those that will remain uniquely human. This distinction serves as a guiding principle for designing AI-native education systems, where greater emphasis is placed on fostering uniquely human perceptual skills, while those within AI's core competencies are carefully integrated. The paper proposes a strategy for equipping AI-natives to collaborate with AI tools in areas where AI currently excels while simultaneously nurturing the skills that complement the perceptive abilities AI may eventually develop. In essence, it develops a holistic spectrum of AI-collaborative and AI-complementary perceptual skills, thereby drafting a blueprint for making education appropriate for Age of Intelligence driven by AI tools.

Keywords: AI-natives · Holistic perception · AI collaboration · AI complementarity · Perceptual skills · AI-native education · Endo-sensory Perceptual Skills · Exo-sensory Perceptual Skills · Meta-sensory Perceptual Skills · Trans-sensory Perceptual Skills

1 Introduction

Different learning theories highlight the importance of developing the perceptual skills of learners. Cognitive theory emphasizes the role of perceptual skills in organizing and categorizing sensory data for meaningful learning and advocates for the use of multimedia tools to engage multiple perceptual skills [1]. Constructivist theory views perceptual skills as integral to building and refining mental models and promotes hands-on activities, such as science experiments, to engage tactile, visual, and auditory perceptions, fostering discovery learning [2]. Sociocultural theory considers perceptual skills essential for interpreting social and cultural cues and promotes group activities and collaborative

M. E. Auer et al. (Eds.): IMCL 2025, LNNS 1954, pp. 69–88, 2026.
https://doi.org/10.1007/978-3-032-24998-2_7

learning, leveraging them to enhance peer-mediated knowledge construction [3]. Experiential learning theory emphasizes learning through experience, reflection, and active experimentation, which calls for the effective use of perceptual skills [4]. Humanist theory views perception as a pathway to self-awareness and meaningful engagement with the world and promotes aesthetic perception nurtured through arts-based learning activities, inspiring creativity and self-expression [5]. The systems theory of learning emphasizes the interaction between individuals and their environmental systems to make learning contextualized [6].

With the advent of AI tools, the understanding of intellectual formation is undergoing a radical paradigm shift. This shift necessitates the careful design and progressive calibration of human perceptual skills to align with the mastery AI tools achieve in this domain. AI-native students must be equipped to collaborate effectively with AI tools and complement their perceptual skills in life and workspaces to avoid redundancy and escape any singularity [7].

Developing a comprehensive understanding of AI-collaborative and AI-complementary perceptual skills is therefore essential for reimagining and redefining education processes suitable for AI natives. This paper seeks to provide an overview of a holistic spectrum of human perceptual skills necessary for AI natives.

2 Interplay of Human and AI Perceptions

Perceptual skills facilitate the perceptual transformation of sensory inputs from human sense organs (perception) before they are processed by the brain for conceptual transformation (conception). As human sensory inputs are themselves influenced by the cognitive functions of the brain, human perceptions are unique. They are shaped by the mindset, value-set, skillset, and life-set, or life experiences and habits, of the perceiver, imparting individual uniqueness to them. For example, repeated exposure to specific sensory stimuli strengthens neural pathways, making certain perceptual interpretations more likely over time.

AI tools are progressively mastering human-like perceptual skills. By leveraging sophisticated algorithms, computational power, and extensive datasets, AI emulates aspects of human perception. AI can also interpret emotional states by analyzing facial expressions, voice intonations, and text through sentiment analysis and emotion detection algorithms [8]. NLP models enhance contextual understanding by analyzing semantic relationships in text and speech [9]. However, AI remains limited by several inherent constraints. It lacks physical embodiment, which restricts its ability to replicate the richness of lived human experiences. It also struggles with tacit cultural knowledge and nuanced situational subtleties that humans naturally comprehend. Moreover, AI is devoid of human-like emotions, self-awareness, intuition, ethics, and aesthetics.

As AI systems progressively master parts of human perceptual skills, reimagining the spectrum of authentically human perceptual skills assumes importance. Further, understanding the interplay between human and AI perceptions becomes essential for redefining appropriate AI-native education.

3 Methodology

This paper aims to provide a comprehensive conceptual framework for the evolution of human perceptual skills. It categorizes these perceptual skills into two broad types: AI-collaborative and AI-complementary. AI-collaborative perceptual skills encompass cognitive abilities in which AI tools have already achieved a high degree of mastery. In such cases, human engagement is most effective when directed toward collaborating with AI to enhance perceptive efficiency. In contrast, AI-complementary perceptual skills comprise abilities that AI tools are unlikely to fully master due to their complexity and the virtually infinite number of parameters required for perfect execution. These skills will remain inherently human and represent core competencies that complement AI-driven perceptual capabilities.

Reviewing available literature, this paper distinguishes between AI-collaborative and AI-complementary perceptual skills. Further research is needed to rigorously validate the distinctions between different perceptual skills drawn in this paper. The primary objective of this study is to develop a broad perspective on human perceptual skills in the era of AI, deriving critical insights for AI-native education.

4 Holistic Spectrum of Human Perceptual Skills

The evolution of AI marks a profound shift in how perceptual skills are understood and applied in learning processes. Learning occurs incrementally through persistent interactions between sensory experiences, accumulated knowledge and skills, the self-definition of learners, and the transcendental references of their holistic growth and development. Accordingly, human perceptual skills may be classified into four categories: Endo-sensory, Exo-sensory, Meta-sensory, and Trans-sensory, reflecting increasingly complex perceptual transformations of sensory inputs before they undergo conceptual processing in the human brain. This progression is also expected to highlight the extent to which AI can replicate or assist in these cognitive transformations.

The first two categories, Endo-sensory and Exo-sensory perceptual skills, are considered AI-collaborative, as they can be effectively exercised in partnership with AI tools, which may eventually replace them. In contrast, Meta-sensory and Trans-sensory perceptual skills are AI-complementary, representing uniquely human competencies that AI can assist with but not completely replace.

Endo-sensory perceptual skills are grounded in the basic human senses: sight, hearing, taste, touch, and smell, whichever sensory organs humans use to interact with and interpret the world around. These skills involve direct sensory inputs, requiring minimal interpretation or transformation. In contrast, Exo-sensory skills extend beyond raw sensory data, incorporating contextual information, accumulated knowledge, and learned skills and abilities. Both of these perceptual skillsets can be augmented by emerging AI tools, justifying their classification as AI-collaborative perceptual skills.

Meta-sensory and Trans-sensory perceptual skills represent deeper layers of human perception informed by the unique perspectives of the perceiver, making them by and large beyond the reach of AI. Meta-sensory skills involve the transformation of sensory inputs through the perceivers' unique perspectives formed by their mindset, value-set, skillset, and life-set, or life experiences and individual habits. They emerge from what humans truly are, their "being" or self-definition as individuals. Trans-sensory skills, on the other hand, refer to the transformation of sensory inputs via the transcendental perspectives of the perceiver, relating to the absolute norms of their growth and development, their "becoming." Both of these perceptual skills will always surpass the perceptive abilities of AI tools and complement them, justifying their classification as AI-complementary perceptual skills.

This taxonomy of perceptual skills, encompassing both AI-collaborative and AI-complementary dimensions, forms the foundation for the intellectual formation of AI natives. AI-collaborative skills empower individuals to effectively leverage AI tools in life and workspaces, while AI-complementary skills foster uniquely human capabilities essential for their AI resilience.

5 AI-Collaborative Perceptual Skills

Many human perceptual skills are now effectively mimicked by AI tools and are expected to be increasingly handled by them in the future. These skills can be categorized as AI-collaborative perceptual skills, highlighting the necessity for AI-native students to collaborate intelligently with AI tools to enhance their efficiencies.

AI-collaborative perceptual skills primarily involve sensory inputs processed through cognitive mechanisms, which AI systems can easily mimic. While humans rely on neural structures shaped by evolutionary biology, AI utilizes algorithmic and statistical methods trained on extensive datasets. Irrespective of that, collaboration between human and AI perceptual capabilities offers transformative opportunities across diverse domains, including healthcare, business, environmental study, and education.

5.1 Endo-Sensory Perceptual Skills

Humans rely on the five senses, namely, sight, hearing, touch, taste, and smell, for perception. Visual perception enables differentiation through the analysis of light, colour, shape, and motion. Photoreceptors in the retina transmit signals to the visual cortex, where attributes such as contrast and spatial arrangement are processed, facilitating object recognition and scene segmentation [10]. The auditory system discerns sound frequencies, rhythms, and spatial locations. Hair cells in the cochlea convert sound waves into neural signals, enabling the differentiation of voices, instruments, and environmental sounds.

Tactile perception involves mechanoreceptors that detect variations in pressure, vibration, and temperature. These inputs are processed in the somatosensory cortex, enabling the differentiation of textures and sensations [11]. The olfactory system identifies thousands of chemical compounds, allowing for nuanced recognition of specific odours. Signals from the olfactory bulb contribute to detailed scent differentiation [12]. Gustatory perception identifies flavours, namely, sweet, sour, salty, bitter, and umami, via taste buds on the tongue, which relay chemical information to the brain for the differentiation of food items and the assessment of their nutritional value [13].

Endo-sensory perceptual skills are fundamental to human interaction with the environment, representing the initial stage of sensory data transformation before they are subjected to higher-level conceptualization. Neural pathways in the human brain are reinforced through repeated exposure to sensory stimuli [14]. For instance, studies on musicians have demonstrated enhanced auditory discrimination, showing how prolonged training modifies sensory perception [15].

With the advent of artificial intelligence (AI), these skills have become increasingly augmented, automated, and analysed, presenting unparalleled opportunities for human-AI collaboration. AI tools have proven highly effective at replicating, and in some cases surpassing, human capabilities for sensing environmental stimuli. Algorithms process massive datasets to identify patterns, correlations, and anomalies.

AI training, however, is limited by the quality and quantity of labelled or unlabelled data, even though it generalizes well within its training scope. Techniques such as gradient descent and backpropagation enable neural networks to iteratively adjust parameters for optimal performance [16]. AI also struggles with tasks outside its training data. Advanced Machine Intelligence (AMI) aims to narrow this gap by aligning machine perceptions more closely with human perceptions [17].

This underscores the need for a symbiotic relationship between the endo-sensory perceptual skills of humans and AI systems, leveraging the strengths of both to achieve greater effectiveness.

1) *Sensory Substitution*

AI systems mimic various dimensions of human perception with remarkable accuracy. Visual perception is replicated using computer vision technologies, particularly convolutional neural networks (CNNs) [18]. These networks are widely applied in image classification and object detection, achieving human-like performance in specific visual tasks.

Auditory perception is emulated through automatic speech recognition (ASR) and audio analysis systems powered by recurrent neural networks (RNNs) and transformers [19]. These technologies enable AI systems to discern speech, process audio signals, and interpret acoustic patterns with increasing precision.

AI systems are also proficient in olfactory detection. AI-driven electronic noses analyse volatile compounds for applications in healthcare, food safety, and environmental monitoring [20]. These systems provide sophisticated tools for identifying chemical signatures and detecting odours in complex environments.

AI systems are capable of rudimentary gustatory detection as well. Electronic tongues utilize AI to detect and classify taste profiles, supporting quality control in the food and beverage industry [21]. These systems analyse chemical properties to ensure consistency and safety in consumables.

Tactile perception is mimicked using haptic sensors that measure force, pressure, and texture [22]. These sensors enable machines to distinguish between smooth, rough, and granular surfaces, adjust grip strength, and replicate thermoreceptor functions, thereby enhancing human-like touch capabilities in robotics and prosthetics.

AI also replicates temporal perception by analysing time-series data and predicting temporal patterns [23]. It detects changes or patterns over time in sensory data, such as rhythms, frequencies, or movements. For instance, AI can identify rhythmic patterns in music or analyse motion in video feeds [24].

2) *Sensory Expansion*

AI systems can augment human perception by operating at scales beyond natural sensory capabilities. For instance, AI-powered microscopes detect minute cellular changes, aiding in the early diagnosis of diseases [25]. AI algorithms enhance ultrasound imaging, significantly improving diagnostic accuracy [26].

AI-based spectrometers analyse chemical compositions with exceptional precision, enabling advancements in material science, pharmacology, and quality control. AI also integrates thermal data to detect heat signatures, contributing to industrial safety, environmental monitoring, and disaster response.

Thus, AI tools can also serve to expand human perceptual skills, improving their precision and efficiency.

3) *Sensory Integration*

Multisensory integration refers to the ability to combine inputs from different sensory modalities into coherent insights, a task in which AI is becoming increasingly proficient.

Humans naturally merge sensory inputs; for example, when watching someone speak, visual cues (e.g., lip movements) and auditory signals (e.g., speech sounds) integrate to enhance comprehension. The superior colliculus of the human brain facilitates the integration of visual and auditory stimuli [27].

Within individual senses, the human brain binds diverse features into unified representations. For instance, in visual perception, attributes such as shape and motion are integrated to recognize objects [28]. These integrative processes extend to interpreting social interactions, where humans combine cues like facial expressions, gestures, and vocal tones to assess emotions and intentions, a vital skill for interpersonal relationships. However, human perception is susceptible to sensory overload and cognitive biases, which can distort interpretations.

AI systems, in contrast, achieve sensory fusion computationally through algorithmic pipelines. Specialized sensors, such as cameras, microphones, and LIDAR, capture specific types of input, with each modality processed by algorithms tailored to the respective data type. For example, autonomous vehicles achieve visual-auditory integration by fusing data from cameras, LIDAR, and radar, enabling real-time situational awareness and safe navigation [29].

Robotic systems integrate tactile sensors with visual inputs to perform delicate tasks, such as microsurgery or assembling electronics [30]. AI tools also combine sound wave detection and thermal imaging to locate survivors in disaster scenarios, significantly enhancing the effectiveness of search and rescue operations [31].

AI systems perform well in sensory integration today. Continued innovation can improve the adaptability and generalization of AI-driven multisensory integration systems.

4) *Pattern Identification*

Humans naturally identify and process patterns using the prefrontal cortex and hippocampus, which are critical for cognitive functions and memory [32].

In contrast, AI systems achieve pattern recognition through machine learning algorithms, such as Support Vector Machines (SVMs) and neural networks. SVMs are well-suited for analysing structured data, while neural networks are particularly effective for unstructured data like images, text, and audio [33]. These algorithms enable AI systems to recognize intricate patterns in datasets, allowing humans to focus on higher-order interpretation and strategic decision-making. For instance, AI-driven financial analysis platforms identify stock market trends and irregularities, helping investors make informed decisions. Similarly, AI tools analyse climate data to detect patterns and predict weather changes with high accuracy, contributing to more reliable forecasting.

Pattern identification also plays a critical role in outcome prediction. Humans rely on heuristics and experience, while AI employs statistical models and machine learning techniques to analyse data and provide probabilistic forecasts. For example, predictive AI tools in healthcare anticipate disease outbreaks, aiding public health officials in planning effective responses. AI systems can also model disease spread and assist with healthcare logistics, such as resource allocation and vaccine distribution.

In industrial settings, AI systems predict equipment failures by analysing sensor data, enabling timely maintenance and reducing downtime. This predictive capability enhances operational efficiency and minimizes costs, demonstrating the transformative potential of AI in pattern recognition across diverse domains.

5) *Anomaly Detection*

Both humans and AI systems possess perceptual skills for detecting anomalies, a critical capability in fields such as security, medicine, and finance. Humans rely on intuition and accumulated experience to identify anomalies [34]. In contrast, AI employs unsupervised learning techniques, such as clustering and autoencoders, to detect irregularities in data patterns [35].

Collaboration between humans and AI tools significantly enhances anomaly detection. For instance, AI assists in understanding spatial relationships between objects using computer vision systems, depth sensors, and other technologies, while humans provide contextual insights to interpret depth and distance. Similarly, humans and AI systems work together in sensory feature differentiation, such as identifying subtle variations in texture, temperature, or pitch. For example, material quality can be assessed through touch, and tonal differences in speech can be discerned. AI tools equipped with haptic feedback systems and acoustic analysis capabilities augment these tasks, delivering higher precision and scalability.

In short, AI has revolutionized endo-sensory perception by extending the boundaries of human sensory capabilities. It has facilitated advancements ranging from basic detection to enhanced precision, multisensory integration, pattern recognition, and anomaly detection. This paves for human AI-collaboration in making endo-sensory perceptual skills more efficient.

5.2 Exo-sensory Perceptual Skills

According to Merleau-Ponty, human perception is not merely the passive reception of sensory information but an active process involving interpretation and understanding [36]. Perceptual skills related to an advanced layer of perception, where sensory inputs are modulated by prior knowledge stored in the brain's memory cells and by hard skills, commonly referred to as "muscle memory", are termed exo-sensory perceptual skills. The prefix "exo" suggests the transformation of sensory inputs by these agents external to them, including the specific socio-cultural and linguistic context of perception.

Exo-sensory perceptions are facilitated by a network of neural structures, including the prefrontal cortex, basal ganglia, and sensory association cortices [11]. These perceptions are deeply intertwined with accumulated individual knowledge and contexts, significantly influencing how individuals understand, interpret, and respond to their environment.

The interaction between sensory inputs and memory has been extensively studied. Bruner and Postman [37] demonstrated that individuals are more likely to recognize objects aligned with their past experiences or expectations, even when confronted with ambiguous stimuli. Similarly, research by Bar [38] highlights how past experiences stored in memory contribute to rapid object recognition by activating related neural pathways in the prefrontal cortex. Baddeley [39] reviewed how stored experiences, knowledge, and skills shape real-time sensory interpretation, underscoring the dynamic nature of perception.

Both humans and AI systems analyse uncertain or incomplete information to make decisions. While humans rely on heuristics, mental shortcuts shaped by past experience, AI employs probabilistic models and reinforcement learning techniques. Kahneman and Tversky [40] provided evidence of how humans use heuristics for decision-making under uncertainty. In contrast, AI systems use Markov Decision Processes (MDPs) and reinforcement learning to navigate uncertainty in dynamic environments [41].

Exo-sensory perceptual skills include:

1) *Contextual Perception*

Contextual perception involves the ability to identify patterns in sensory inputs by drawing on prior knowledge and experiences. This process enables humans to consider the broader context and nuances of a situation, emphasizing the importance of integrating contextual knowledge with sensory data [42].

AI tools have made significant advancements in contextual adaptation, particularly in structured environments. However, trained human agents are often required to interpret AI-generated results within a given context. For example, AI systems can analyse diagnostic images such as X-rays or MRIs to identify pathological patterns. Nevertheless, a physician's prior knowledge and contextual awareness are crucial for interpreting subtle patterns that AI systems may overlook.

This dynamic applies similarly to other fields, such as linguistic analysis and weather prediction using AI tools. While AI excels in identifying structured patterns within predefined parameters, it struggles with ambiguous or novel contexts that require human judgment informed by prior experiences and expertise. This limitation highlights the complementary roles of humans in contexts where AI's capabilities are insufficient.

2) *Dynamic Perception*

Dynamic adaptation involves the real-time adjustment of sensory inputs based on evolving contexts, requiring significant flexibility. Humans rely on the heuristics they have already developed or to make rapid and informed decisions in dynamic situations.

AI systems have also developed mechanisms for dynamically integrating sensory inputs. For example, self-driving vehicles adjust to changing road conditions by integrating visual, auditory, and environmental data. Despite significant advancements, these systems occasionally struggle with unpredictable human behaviours that experienced drivers can anticipate and adapt to. Similarly, AI tools like Google Translate adapt to spoken language nuances in real time but often falter when handling idiomatic or region-specific expressions [43].

These examples illustrate that while AI systems excel in structured environments, they often lack the intuitive flexibility and foresight demonstrated by humans, limiting their effectiveness in unstructured or unpredictable situations.

3) *Societal Perception*

Societal perception refers to the human ability to understand sensory information within societal contexts [44]. It integrates nuanced understanding, tacit knowledge, and past experience, all of which are essential for navigating complex human interactions and social dynamics [45].

This perception merges tacit knowledge with social underpinnings. It is critical for tasks such as conflict resolution, storytelling, and communication. For example, societal values vary significantly, influencing how individuals perceive actions, symbols, and social norms. While AI tools excel at processing data and identifying patterns, they lack the lived experience and societal nuance necessary to fully understand context.

In legal systems, for instance, judges interpret laws through ethos of society. These tasks require contextual understanding and social competencies, capabilities that often remain beyond the reach of AI. Without exhaustive datasets and sophisticated training methods, AI systems struggle to adapt to the variability and subtlety of societal contexts [46].

However, AI tools can augment human perceptions with data-driven insights into societal dynamics, making AI-collaboration still meaningful. Hence, it is imperative for AI-natives to continually refine and improve their societal competencies through educational processes in collaboration with AI tools to hone their decision-making skills.

4) *Cultural Perception*

Cultural perception involves the transformation of sensory inputs through the lens of their cultural experiences. This ability enables humans to decode insights that AI tools often misinterpret. For instance, while AI tools can analyse the artistic styles and compositions of a work, the concrete cultural background of the perceiver significantly shapes and enriches their interpretation.

A similar challenge arises in perceiving sensory inputs related to cultural archetypes. An in-depth understanding of these archetypes is developed through lived experiences and immersion within a community of practice, something that remains formidable for AI tools to achieve [47].

Human agents play an essential role in providing nuanced cultural interpretations of sensory information. In practice, AI and human agents can work collaboratively. While AI can interpret user sentiment and tailor responses, human agents excel in understanding complex cultural nuances, addressing them with greater depth and effectiveness. Culturally agnostic AI models are a myth [48].

5) *Linguistic Perception*

Humans and AI systems share the ability to process and interpret spoken language. In humans, the superior temporal gyrus plays a critical role in language processing, enabling the comprehension of spoken language [49]. AI achieves similar functionality through natural language processing (NLP) technologies. Tools such as Google Speech-to-Text and Siri leverage deep learning models to achieve near-human accuracy in speech recognition tasks [50].

While affective computing can identify certain tones in speech, it often struggles with subtleties such as sarcasm, irony, or cultural idioms. Similarly, AI-driven chatbots and virtual assistants provide contextual responses but frequently fail to adapt to subtle shifts in user intent or conversational style. Moreover, AI models can inadvertently reinforce biases present in their training data, resulting in outputs that perpetuate stereotypes or discriminatory language.

Human expertise is crucial for identifying and mitigating these biases, ensuring fair and equitable language processing outcomes [51]. Human intervention also provides necessary insights to ensure that AI-generated language aligns with societal norms and values [52]. Additionally, continuous human input is essential for updating AI systems with current linguistic trends, styles, and usages, thereby maintaining their relevance and accuracy in language processing [53].

AI systems have demonstrated remarkable potential in augmenting exo-sensory perceptual skills, particularly within well-defined and structured contexts. However, the inherent subjectivity, cultural diversity, and adaptability of human perception remain significant barriers to fully replicating human capabilities. At present, human prior knowledge and specialized skills play a critical role in shaping exo-sensory perceptual skills, enabling a level of adaptability and depth that AI systems have yet to achieve. This gap underscores the existing potential for deeper collaboration between humans and AI.

6 AI-Complementary Perceptual Skills

While AI tools have made significant strides in mimicking AI-collaborative perceptual skills, some authentic human perspectives will remain beyond the full reach of AI, due to the presence of irreducible dimensions, as described in this section. These abilities are referred to as AI-complementary perceptual skills, which humans can leverage to complement and thus enhance the efficiency of AI tools. The design of educational processes for AI-natives must prioritize the development of such AI-complementary perceptual skills, preparing them to thrive in the AI age.

AI-complementary perceptual skills or perspectives can be categorized into two groups, arranged in order of progressively decreasing reliance on AI tools in transforming sensory data into perceptions:

- Meta-sensory perceptual skills: These are perspectives shaped by the self-definition or individuality of the perceiver. These skills transform sensory data based on the unique personal identity or "being" of the perceiver, characterized by an individual's mindset, value-set, skillset, and life-set or habits of life.

- Trans-sensory perceptual skills: These are perspectives shaped by the transcendental ideals of the perceiver. These skills transform sensory data based on the unique transcendental drives of the perceiver that set the stage for their further development or "becoming".

AI-complementary perceptual skills can augment well with the perceptual abilities of AI tools, synergistically complementing them in acquiring perfection in perception. Therefore, an education system that nurtures AI-complementary perceptual skills can help AI-natives contribute meaningfully to AI-driven workspaces while remaining authentically human.

6.1 Meta-Sensory Perceptual Skills

Meta-sensory perceptions are deeply rooted in the individual perspectives of the perceiver and are characterized by their mindset, value-set, skillset, and life-set or habits of life. These skills impart an embodied, temporal, emotional, personal and mutual transformation to the sensory inputs. They are characterized by the integration of sensory information with the perceiver's personal identity.

A network of brain structures, including the prefrontal cortex, anterior cingulate cortex, and insula, collectively supports these higher-order perceptions [11]. While AI tools can mimic certain aspects of meta-sensory perceptual skills using advanced algorithms, their inability to embody human-like identity limits their capacity to fully replicate these skills.

1) *Embodied Perception*

Embodied perception refers to the body's inherent ability to perceive stimuli related to touch, position, and internal physiological states, collectively defining somatosensation. This encompasses the automatic and instinctual processing of sensory information from within the body and the immediate environment. According to Johnson [54], somatosensation involves integrating sensory experiences with movement and spatial awareness. A network of brain regions, including the brainstem, thalamus, and primary sensory cortices, supports somatosensation [11]. These regions process sensory information hierarchically and distributively, with early sensory processing occurring in primary sensory areas and higher-level processing in association cortices.

At its core, human perception is rooted in somatosensation, expressed through three key components:

- Proprioception: The perception of body position, movement, and tension [55].
- Interoception: The perception of internal bodily states, such as hunger, thirst, and pain [56].
- Exteroception: The perception of external sensory information, such as light, sound, and touch [57].

AI systems are rapidly advancing in simulating various aspects of somatosensation to improve autonomous robotics. Proprioceptive sensors help robots understand and predict their movements and positions in space, mimicking human proprioception [58]. The integration of interoceptive signals is critical for developing AI systems that better interpret human emotions and physiological states. Additionally, soft robots equipped with sensorized skins can perceive external pressures and adapt their actions accordingly, mimicking human exteroceptive capabilities [59].

Somatosensation provides humans with an intuitive and meaningful connection to their environment via embodied or physical perception. In contrast, AI-powered robots, despite their advanced sensory capabilities, lack the subjective experience of physicality.

Educational activities that enhance physical intuition, such as dance or hands-on craftsmanship, play a crucial role in fostering embodied intelligence. In general, integrating any kind of physical activity into AI-native education is essential for developing somatosensation.

2) *Temporal Perception*

Human perception is inherently temporal, involving the ability to understand the flow of time and perceive events within a temporal framework [60]. Temporal perception integrates episodic memory, temporal binding, and prospection, supporting the development of long-term plans and strategies as well as the construction of impactful narratives and stories.

Episodic memory allows individuals to mentally revisit and re-experience past events, enabling the reconstruction of past experiences with rich context and emotional colour [61].

Temporal binding refers to the understanding of time, encompassing the ability to process temporal order, duration, and simultaneity. This capability facilitates cohesive perceptions of events and enables the understanding of sequences and causality [62]. Humans perceive historical events through cultural and temporal lenses, incorporating their own personal and critical reflections.

Prospection involves futuristic perspectives, supporting planning, prediction, and preparation for future events [63]. It enables humans to imagine potential scenarios, facilitating effective decision-making and goal-setting [64].

While AI excels at storing and retrieving large-scale datasets, it relies on human agents to provide critical context and assess data relevance. AI systems optimize workflows, generating variations, and testing potential scenarios with efficiency and precision. They excel at processing time-series data and identifying temporal patterns. However, they lack the subjective and intuitive experience of time and history that humans possess. Cultivating foresight and temporal awareness in humans should, therefore, be an essential component of AI-native education, equipping students with this critical AI-complementary temporal perceptual skills.

3) *Emotional Perception*

Research by Vuilleumier [65] demonstrates how emotional salience, informed by past experiences, heightens attentional focus and perception through amygdala activation. Humans inherently filter sensory inputs through the lens of emotions [66]. They integrate emotions with sensory information, which can be inferred from facial expressions [67].

AI has also developed tools for facial recognition and sentiment analysis through affective computing [8]. For example, AI-driven emotion recognition systems are increasingly employed in diagnosing and monitoring mental health conditions. These tools analyse emotional cues to support therapy plans and facilitate the early detection of disorders. Similarly, wearable emotion recognition devices assist individuals with autism by analysing facial expressions and body language, providing real-time feedback to help them interpret social cues more effectively.

However, human expertise and prior knowledge play a critical role in meaningfully integrating the emotional cues identified by AI systems.

Emotional integration remains a key domain where human perceptual skills can complement those of AI systems. Authentic emotional self remains core-competence of an individual, who alone can ensure effective generation and management of emotional responses. Hence, development of emotional intelligence assumes vital importance in Ai-native education.

4) *Personal Perception*

Personal perception is personality-driven, shaped by the unique mindset, values, attitudes, seasoned with life experiences of each perceiver. Sensory information is filtered and interpreted through the lens of personal values, experiences, and biases [68].

Personality traits and life experiences heavily influence individual perception. Two individuals may perceive the same sensory input differently based on their unique identities and traits. For example, resilient individuals are more likely to frame challenges positively, viewing them as opportunities rather than threats, an outlook linked to traits like optimism and self-efficacy [69].

Similarly, writers and artists often draw upon their distinctive personalities and life experiences to create deeply personal works that emotionally resonate with others. This subjective and experiential basis of creative expression makes personality-driven perception particularly challenging for AI to replicate.

Personal perception is enhanced by self-awareness, which allows individuals to recognize their biases and personal filters. A well-developed sense of identity and mindfulness helps reduce distortion caused by personal prejudices [70].

While AI can identify patterns of cognitive biases in decision-making, it lacks the capacity for self-reflection, a fundamental aspect of human perception. Thus, development of authentic personality in students assumes priority in designing an AI-native education.

5) *Mutual Perception*

Mutual perception involves interpreting sensory inputs through a synergy of different perspectives of perceivers, rooted in empathy and emotional intelligence. It is driven by the compassionate and altruistic perspectives of the perceivers and is linked to neurological processes such as the activation of mirror neurons, which are biologically inherent to humans [71].

Perspective-taking, a key component of mutual perception, is influenced by traits such as openness to interpersonal experiences and engagements. Although AI systems can simulate perspective-taking using algorithms, humans provide deeper insights by integrating synergistic nuances that go beyond computational capabilities.

Mutual perception also includes the ability of individuals to create identities in relation with other individuals. For example, head of a family incorporates in himself/herself the perspectives of the family or head of a state represents the perspectives of every citizen of the state. Such identities based on the mutuality of individuals can also colour the perspectives with which an individual develops the mental picture of reality.

While AI tools can simulate emotional states, they lack genuine empathy, which requires mentally placing oneself in another's situation. Consequently, mutual perception relies on human intervention to enrich AI judgments with empathy, ethical reasoning, and contextual depth. Hence, development of skills associated with mutual perception assumes importance in the definition of an AI-native education.

Meta-sensory perceptual skills represent an advanced layer of human perception, where sensory inputs are filtered and interpreted through the embodiment, historicity and unique identity of perceivers, defined by their mindset, value-set, attitudes and interpersonal identities. These skills rely on unique capacities of individuals related to mindfulness, emotional intelligence, and synergy, thus forming a domain that remains distinctly human. While AI systems can enhance meta-sensory perceptual skills by providing tools and data to support them, the essence of these skills lies in the ability of an individual to interpret sensory information from a perspective unique to the perceiver.

6.2 Trans-Sensory Perceptual Skills

Trans-sensory perception arises from transcendental perspectives rooted in humanity's collective and an individual's personal quest for transcendentals, including Truth, Goodness, and Beauty, as expressed through philosophy, ethics, and aesthetics, respectively. These transcendentals cannot be reduced to a finite number of parameters such that they can be fully exhausted by AI tools. Transcendental perspectives are capable of informing and transforming sensory information, helping humans transcend the limitations of physical senses to access deeper dimensions of reality.

At their most profound level, trans-sensory perceptual skills are characterized by phenomena like cosmic consciousness, the experience of interconnectedness with all existence [72]; peak experiences associated with self-actualization and transcendence [73]; and altered states of consciousness [74]. While the mechanisms behind them are not fully understood, research suggests that neuroplasticity and neurotransmitters, such as serotonin, dopamine, and endorphins, play an integral role in such transformative processes [75].

Due to their deeply metaphysical nature, trans-sensory perceptual skills are entirely beyond the simulation capabilities of AI tools. These tools lack the intrinsic capacity to engage with the transcendental dimensions of human perception. Thus, nurturing trans-sensory perceptions can serve as the cornerstone of AI-native education, imparting AI resilience.

1) *Philosophical Perception*

The sense of truth involves the ability to perceive and resonate with fundamental principles concerning reality in its totality. It encompasses the capacity to understand abstract truths and engage with philosophical questions about existence. Nagel [76] describes this as a "view from nowhere," referring to an idealized objective perspective, a viewpoint that transcends individual biases, emotions, and personal experiences.

The human search for absolute truth has given rise to philosophy, which seeks the ultimate truth about existence, morality, and the universe through rationality, intuition, and deeply personal quests for understanding.

While AI systems excel at processing information logically and efficiently, they cannot grasp the fullness of truth, as they lack the subjective self-awareness required for exploring metaphysical or existential questions. Thus, fostering philosophical perspectives in students is a crucial component of AI-native education to achieve AI resilience.

2) *Ethical Perception*

The sense of goodness reflects the human ability to perceive and evaluate the ethical dimensions of actions and decisions. It refers to the capacity to distinguish between right and wrong, or good and bad, based on a categorical imperative, a universal principle of action rooted in human autonomy [77]. This sense of goodness arises from a profound commitment to moral integrity, fairness, and equity that transcends self-interest of the perceiver.

While AI can follow programmed ethical guidelines and model scenarios based on predefined frameworks, it cannot intuitively navigate complex moral and ethical dilemmas. Ethical reasoning requires subjective judgment shaped by cultural, historical, and situational factors, elements that AI cannot internalize.

Consequently, humans must strengthen their moral reasoning and ethical leadership to serve as stewards of AI applications. They are responsible for designing, implementing, and monitoring AI systems to ensure alignment with ethical principles and to mitigate risks such as bias or misuse, particularly in sensitive fields like healthcare and criminal justice [78].

3) *Aesthetic Perception*

The sense of beauty involves the perception of harmony, proportion, and aesthetic value in sensory experiences [79]. It encompasses the appreciation of natural elements, such as a sunset, a forest, or the ocean, which evoke profound emotions of awe and wonder. It also includes the creation of something beautiful through inspiration, emotional depth, and personal vision. The sense of beauty represents a deeply personal yet universal perception of aesthetic harmony, encompassing art, nature, and human expression.

Aesthetic perception elicits a deeply human response that goes beyond the capacity of AI to experience or replicate. While AI tools, such as generative adversarial networks (GANs), can produce aesthetically pleasing visuals, they lack the subjective engagement with beauty and the emotional depth that characterize human creativity.

Hence, education of AI-natives must be anchored on developing aesthetic perceptual skills to help them complement well with AI tools. By nurturing this uniquely human capacity, individuals can confidently engage in endeavours that enrich cultural and creative tastes of humankind in an increasingly AI-driven world.

4) *Existential Perception*

Existential perception connects humans to their unique sense of rootedness in the universe. It enables them to perceive themselves as part of an interdependent and interconnected reality within the vast cosmos. It imparts a systemic view of life, offering a unifying vision to understand and synthesize dynamic, interconnected systems by integrating diverse inputs to evaluate systemic impacts [80].

AI systems process inputs objectively but lack the existential depth required to replicate deeply existential perceptions. They cannot engage with questions about their own existence or purpose.

Thus, developing skills related to existential perception equips students to oversee AI-driven processes and address their unintended existential consequences and reigns supreme in the definition of AI-native education.

5) *Spiritual Perception*

Spirituality refers to the recognition of the sacred dimensions of existence [81]. It connects humans to the deepest purpose and meaning of life, guiding ethical and cultural evolution. Spirituality can radically transform human perception as it deals with profound inner connections, fostering a sense of love and compassion that transcends biological or transactional interactions [82]. Spiritual perception often emphasizes love as a transcendental value, promoting unity and universal harmony. It also inspires acts of compassion rooted in a profound sense of altruism.

AI, by contrast, lacks the capacity for spiritual experiences due to the absence of consciousness and self-reflection. While AI tools, such as chatbots, can simulate empathy to some degree, they cannot draw on spiritual awareness. Therefore, fostering spiritual and reflective practices in students is essential for preparing them for the AI age.

In short, though AI tools can complement human efforts to engage with trans-sensory perceptions of reality by providing data, tools, and platforms that support the development of transcendental perspectives, they cannot replicate the deeply transcendental, subjective, and spiritual dimensions of such perceptions. By cultivating trans-sensory perceptual skills, AI-natives can enhance their capacity to work synergistically with AI systems while retaining the unique qualities that define humanity. Hence, the development of trans-sensory perceptual skills remain vital for preparing students to thrive in an AI-driven future.

7 Limitations and Outlook

This paper aims to develop a conceptual framework for describing suitable perceptual skills for AI-natives. Its scope is limited to distinguishing between AI-collaborative and AI-complementary aspects of perceptual skills, which helps define a holistic spectrum of perceptual skills for AI-natives. It does not deliberate on the current state of AI usage in educational contexts or assess its impact. Nor does it describe curricular processes or classroom scenarios for developing perceptual skills or validate its findings against generated data.

As a continuation of this work, models for integrating the holistic spectrum of perceptual skills into various curricular processes will be developed. Furthermore, quantitative metrics to measure the development of perceptual skills in students, based on a pilot implementation, will be defined. This will facilitate the creation of case studies that offer practical insights and recommendations for effectively integrating development of perceptual skills into AI-native education.

8 Conclusion

To navigate the complexities of the AI age, AI natives must master a holistic spectrum of perceptual skills encompassing both AI-collaborative and AI-complementary dimensions. AI-collaborative perceptual skills represent areas where AI tools are progressively achieving mastery, enabling humans to enhance efficiency through collaboration with these technologies. This category includes endo-sensory perceptual skills, which are directly linked to human sensory organs (what perceivers "have"), and exo-sensory perceptual skills, which involve sensory inputs shaped by prior knowledge and accumulated hard skills (what perceivers "know"). In contrast, AI-complementary perceptual skills represent areas where AI tools are unlikely to achieve expertise, thereby preserving them as core human competencies. These include meta-sensory perceptual skills, which involve sensory inputs influenced by the perceiver's mindset, value-set, skillset, and life-set (what perceivers "are"), and trans-sensory perceptual skills, which transcend sensory inputs to engage with the transcendental dimensions of human perception, such as Truth, Goodness, and Beauty (what perceivers "become").

While AI-collaborative perceptual skills are necessary for equipping students to work effectively in collaboration with AI tools, mastery of AI-complementary perceptual skills is the sufficient condition for preserving and advancing uniquely human competencies. Every aspect of AI-native education must take both categories of skills into active consideration and restructure its framework to prioritize AI-complementary perceptual skills over AI-collaborative ones. This paradigm shift is essential, as current educational practices often emphasize skills in which AI tools are progressively gaining mastery, exposing AI natives to the danger of replacement by AI tools.

In summary, this paper calls for a reprioritization of training in perceptual skills in response to the rapid expansion of AI capabilities into areas traditionally considered core human competencies. By fostering an education system rooted in a holistic spectrum of AI-collaborative and AI-complementary perceptual skills, with a focus on AI-complementary perceptual skills, AI-native education can ensure that human perceptual skills remain relevant in the workspaces and life-spaces of an AI-driven world.

References

1. Neisser, U.: Cognitive Psychology. Appleton-Century-Crofts, New York (1967)
2. Piaget, J.: The Psychology of Intelligence. Totowa, NJ: Littlefield, Adams (1966)
3. Vygotsky, L.S.: Mind in Society: The Development of Higher Psychological Processes. Harvard Univ. Press, Cambridge, MA (1978)
4. Kolb, D.A.: Experiential Learning: Experience as the Source of Learning and Development. Prentice Hall, Englewood Cliffs, NJ (1984)
5. Rogers, C.R.: Freedom to learn: a view of what education might become. Columbus, OH: Charles Merrill (1969)
6. Bronfenbrenner, U.: The Ecology of Human Development: Experiments by Nature and Design. Harvard Univ. Press, Cambridge, MA (1979)
7. Panthalookaran, V.: Entrepreneurial Intellection: Shades of Black & White. Pallikkutam Publications, Kochi (2022)
8. Cambria, E., Das, D., Bandyopadhyay, S., Feraco, A.: Affective computing and sentiment analysis. Springer (2017)

9. Devlin, J.: Bert: Pre-training of deep bidirectional transformers for language understanding (2018). arXiv preprint arXiv:1810.04805
10. Zeki, S.: A vision of the brain. Blackwell (1993)
11. Kandel, E.R., et al.: Principles of Neural Science, vol. 4. McGraw-Hill, New York (2000)
12. Buck, L., Axel, R.: A novel multigene family may encode odorant receptors: A molecular basis for odor recognition. Cell **65**(1), 175–187 (1991)
13. Chandrashekar, J., et al.: The receptors and cells for mammalian taste. Nature **444**(7117), 288–294 (2006)
14. Hebb, D.O.: The organization of behavior: a neuropsychological theory. Wiley (1949)
15. Wan, C.Y., Schlaug, G.: Music-making as a tool for promoting brain plasticity across the life span. Neuroscientist **16**(5), 566–577 (2010)
16. LeCun, Y., Bengio, Y., Hinton, G.: Deep learning. Nature **521**(7553), 436–444 (2015)
17. Felin, T., Holweg, M.: Theory is all you need: AI, human cognition, and causal reasoning. Strategy Sci. (2024)
18. Krizhevsky, A., Sutskever, I., Hinton, G.E.: ImageNet classification with deep convolutional neural networks. Adv. Neural. Inf. Process. Syst. **25**, 1097–1105 (2012)
19. Hinton, G., et al.: Deep neural networks for acoustic modeling in speech recognition: the shared views of four research groups. IEEE Signal Process. Mag. **29**(6), 82–97 (2012)
20. Wilson, A.D., Baietto, M.: Applications and advances in electronic-nose technologies. Sensors **9**(7), 5099–5148 (2009)
21. Das, S., et al.: Robust chemical analysis with graphene chemosensors and machine learning. Nature **616**(7956), 123–129 (2024)
22. Dahiya, R.S., Metta, G., Valle, M., Sandini, G.: Tactile sensing—From humans to humanoids. IEEE Trans. Robot. **26**(1), 1–20 (2010)
23. Graves, A., Mohamed, A.R., and Hinton, G.: Speech recognition with deep recurrent neural networks. In: Proc. IEEE Int. Conf. Acoust., Speech Signal Process. (ICASSP), pp. 6645–6649 (2013)
24. Wang, Y., Ostermann, J., Zhang, Y.Q.: Video Processing and Communications. Prentice Hall, Upper Saddle River, NJ (2002)
25. Litjens, G., et al.: A survey on deep learning in medical image analysis. Med. Image Anal. **42**, 60–88 (2017)
26. Huang, Q., Zhang, F., Li, X.: Machine learning in ultrasound computer-aided diagnostic systems: a survey. Biomed. Signal Process. Control **56**, 101675 (2020)
27. Stein, B.E., Stanford, T.R.: Multisensory integration: current issues from the perspective of the single neuron. Nat. Rev. Neurosci. **9**(4), 255–266 (2008)
28. Treisman, A.: The binding problem. Curr. Opin. Neurobiol. **6**(1), 171–178 (1996)
29. Chen, X., Ma, H., Wan, J., Li, B., Xia, T.: Multi-view 3D object detection network for autonomous driving. In: Proc. IEEE Conf. Comput. Vis. Pattern Recognit., pp. 1907–1915 (2017)
30. Kappassov, Z., Corrales, J.A., Perdereau, V.: Tactile sensing in dexterous robot hands—Review. Robot. Auton. Syst. **74**, 195–220 (2015)
31. Yeom, S.: Thermal image tracking for search and rescue missions with a drone. Drones **8**(2), 53 (2024)
32. Cohen, N.J., Eichenbaum, H.: Memory, Amnesia, and the Hippocampal System. MIT Press (1993)
33. Hastie, T., Tibshirani, R., Friedman, J.: The elements of statistical learning: data mining, inference, and prediction. Springer (2009)
34. Kahneman, D., Klein, G.: Conditions for intuitive expertise: a failure to disagree. Am. Psychol. **64**(6), 515–526 (2009)
35. Chandola, V., Banerjee, A., Kumar, V.: Anomaly detection: a survey. ACM Comput. Surv. **41**(3), 1–58 (2009)

36. Merleau-Ponty, M.: Phenomenology of perception, C. Smith, Trans. London, Routledge (1945)
37. Bruner, J.S., Postman, L.: On the perception of incongruity: a paradigm. J. Pers. **18**(2), 206–223 (1949)
38. Bar, M.: The proactive brain: using analogies and associations to generate predictions. Trends Cogn. Sci. **11**(7), 280–289 (2007)
39. Baddeley, A.D.: Working memory: theories, models, and controversies. Annu. Rev. Psychol. **63**(1), 1–29 (2012)
40. Kahneman, D., Tversky, A.: Prospect theory: an analysis of decision under risk. Econometrica **47**(2), 263–291 (1979)
41. Sutton, R.S., Barto, A.G.: Reinforcement Learning: An Introduction. MIT Press (2018)
42. Barsalou, L.W.: Grounded cognition. Annu. Rev. Psychol. **59**, 617–645 (2008)
43. Wu, Y., et al.: Google's neural machine translation system: Bridging the gap between human and machine translation (2016). arXiv preprint arXiv:1609.08144
44. Bar, M.: Visual objects in context. Nat. Rev. Neurosci. **5**(8), 617–629 (2004)
45. Adolphs, R.: The social brain: Neural basis of social knowledge. Annu. Rev. Psychol. **60**, 693–716 (2009)
46. Müller, V.C.: Ethics of artificial intelligence and robotics. In: The Stanford Encyclopedia of Philosophy, E. N. Zalta, Ed., Fall 2020 Edition. Stanford Univ. (2020)
47. Prabhakaran, V., Qadri, R., Hutchinson, B.: Cultural incongruencies in artificial intelligence (2022). arXiv preprint arXiv:2211.13069
48. Cetinic, E.: The myth of culturally agnostic AI models (2022). arXiv preprint arXiv:2211.15271
49. Hickok, G., Poeppel, D.: The cortical organization of speech processing. Nat. Rev. Neurosci. **8**(5), 393–402 (2007)
50. Xiong, W., et al.: Achieving human parity in conversational speech recognition (2016). arXiv preprint arXiv:1610.05256
51. Feng, S., Park, C.Y., Liu, Y., Tsvetkov, Y.: From pretraining data to language models to downstream tasks: tracking the trails of political biases leading to unfair NLP models (2023). arXiv preprint arXiv:2305.08283
52. Buddemeyer, A., Walker, E., Alikhani, M.: Words of Wisdom: representational Harms in Learning from AI Communication (2021). arXiv preprint arXiv:2111.08581
53. Smith, G., et al.: Standard language ideology in AI-generated language (2024). arXiv preprint arXiv:2406.08726
54. Johnson, M.: Philosophy in the flesh: the embodied mind and its challenge to western thought. Basic Books (1999)
55. Sherrington, C.S.: The Integrative Action of the Nervous System. Routledge, London (2023)
56. Craig, A.D.: How do you feel? An interoceptive moment with your neurobiological self. Princeton Univ. Press (2015)
57. Gibson, J.J.: The Ecological Approach to Visual Perception. Houghton Mifflin, Boston, MA (1979)
58. Jiang, S., Zhang, J., Wong, L.: Robot body schema learning from full-body extero/proprioception sensors. (2024). arXiv preprint arXiv:2402.18675
59. Soter, G., Conn, A., Hauser, H., Rossiter, J.: Bodily aware soft robots: integration of proprioceptive and exteroceptive sensors. In: Proc. IEEE Int. Conf. Robot. Autom. (ICRA), pp. 2448–2453 (2018)
60. Wittmann, M.: The inner experience of time. Philos. Trans. R. Soc. B: Biol. Sci. **364**(1525), 1955–1967 (2009)
61. Tulving, E.: Episodic memory: from mind to brain. Annu. Rev. Psychol. **53**(1), 1–25 (2002)
62. Zacks, J.M., Tversky, B.: Event structure in perception and conception. Psychol. Bull. **127**(1), 3–21 (2001)

63. Gilbert, D.T., Wilson, T.D.: Prospection: experiencing the future. Science **317**(5843), 1351–1354 (2007)
64. Suddendorf, T., Corballis, M.C.: The evolution of foresight: what is mental time travel, and is it unique to humans? Behav. Brain Sci. **30**(3), 299–313 (2007)
65. Vuilleumier, P.: How brains beware: neural mechanisms of emotional attention. Trends Cogn. Sci. **9**(12), 585–594 (2005)
66. Goleman, D.: Emotional Intelligence: why It Can Matter More Than IQ. Bloomsbury Publishing (2020)
67. Adolphs, R.: Recognizing emotion from facial expressions: psychological and neurological mechanisms. Behav. Cogn. Neurosci. Rev. **1**(1), 21–62 (2002)
68. Fodor, J.A.: The Modularity of Mind. MIT Press (1983)
69. Connor, K.M., Davidson, J.R.T.: Development of a new resilience scale: the Connor-Davidson Resilience Scale (CD-RISC). Depress. Anxiety **18**(2), 76–82 (2003)
70. Brown, K.W., Ryan, R.M., Creswell, J.D.: Mindfulness: theoretical foundations and evidence for its salutary effects. Psychol. Inq. **18**(4), 211–237 (2007)
71. Decety, J., Jackson, P.L.: The functional architecture of human empathy. Behav. Cogn. Neurosci. Rev. **3**(2), 71–100 (2004)
72. Bucke, R.M.: Cosmic consciousness: a study in the evolution of the human mind. EP Dutton (1960)
73. Maslow, A.H.: Religions, values, and peak-experiences. Ohio State Univ. Press (1964)
74. Tart, C.T.: Altered States of Consciousness. Anchor Books (1972)
75. Cahill, R.E., Stevenson, J.H.: Neuroplasticity and neurotransmitter dynamics in sensory integration. Front. Neurosci. **14**, 1–15 (2020)
76. Nagel, T.: The View from Nowhere. Oxford Univ. Press (1989)
77. Kant, I.: Critique of Practical Reason, T. K. Abbott, Trans. Dover Publications (2004)
78. Floridi, L., Cowls, J.: A unified framework of five principles for AI in society. Harvard Data Sci. Rev. **1**(1) (2019)
79. Leder, H., et al.: A model of aesthetic appreciation and aesthetic judgments. Br. J. Psychol. **95**(4), 489–508 (2004)
80. Capra, F., Luisi, P.L.: The Systems View of Life: A Unifying Vision. Cambridge Univ. Press (2014)
81. James, W.: The Varieties of religious experience: a study in human nature. Mod. Libr. (1902)
82. Tillich, P.: The Courage to Be. Yale Univ. Press (1952)

Psychometric Test for Intrinsic Learning Motivations of AI-Natives

Varghese Panthalookaran[1(✉)], Christina Sotiriadou[2], and Bernd Zinn[2]

[1] Department of Basic Sciences and Humanities, Rajagiri School of Engineering and Technology, Kochi 682039, India
panthalookarancmi@gmail.com, varghese@rajagiritech.ac.in

[2] Institute of Educational Science, Department of Vocational Education Focused On Teaching Technology, University of Stuttgart, 70174 Stuttgart, Germany

Abstract. Students of the AI age are required to navigate a volatile, uncertain, complex, and ambiguous (VUCA) world. To equip themselves, they must adopt conducive learning styles that are autonomous, continuous, venturesome, and flexible. Education driven by intrinsic learning motivations is found to consolidate these learning styles in new-generation students [1, 2]. This paper attempts to identify a broad spectrum of suitable intrinsic learning motivations that support the education of students in the AI age. The paper ventures into the construction of a psychometric tool sensitized to the growth and development of intrinsic learning motivations in students. The tool development follows typical statistical processes in social sciences. Statistical procedures, including inter-rater reliability assessment, exploratory factor analysis, and confirmatory factor analysis, are employed on a set of items developed for this purpose. The developed tool is also applied to capture the growth of intrinsic learning motivations in students participating in an academic program. The responses of students at the beginning and towards the end of the educational intervention are used to assess the growth profiles of intrinsic learning motivations. The results suggest that the psychometric tool thus developed is sufficiently sensitive to capture the gradual growth in the intrinsic learning motivations of students. They also pave way for employing the psychometric tool to improve the efficacy of academic programs in instilling intrinsic learning motivations in learners.

Keywords: AI-native Education · Psychometric test · Learning motivations · Intrinsic learning motivations · Joy of knowledge creation · Joy of self-realization · Joy of self-expression · Joy of playful synergy · Joy of selfless service

1 *I*ntroduction

In the AI age, careers are quickly transforming into either entrepreneurial or intrapreneurial owing to the volatility, uncertainty, complexity and ambiguity (VUCA) associated with it. To equip themselves for this emerging career scenario, new-generation students must develop an entrepreneurial mindset grounded in autonomous, lifelong,

M. E. Auer et al. (Eds.): IMCL 2025, LNNS 1954, pp. 89–110, 2026.
https://doi.org/10.1007/978-3-032-24998-2_8

venturesome, and flexible learning styles. Educational processes driven by intrinsic learning motivations of the students can help achieve this objective of modern-day education [1]. The current paper aims to develop a psychometric tool to measure the efficacy of academic programs in generating intrinsic learning motivations in students.

Educational processes thrive on student motivations. [3] provided an extensive review of research on motivations in educational contexts. [4] examined the relationship between different types of motivation and academic achievement. [5] developed a comprehensive textbook on motivations in educational settings. [6] provided a detailed overview of research on motivation in education.

Student motivations can be either extrinsic or intrinsic. Intrinsic learning motivations are internal factors that drive individuals to engage in education for the sheer joy of learning. Unlike extrinsic motivations, which anticipate external triggers, intrinsic motivations stem from a deep-seated desire to engage with the study. [7] defined relevant intrinsic and extrinsic motivations of students, which often interact and intersect. [8] showed how external rewards can affect intrinsic motivation. [9] concluded that while extrinsic rewards can temporarily boost performance, they can undermine the drive for learning in the long run. Further, external rewards can incentivize students to perform well in simple mechanistic tasks but can lead to poorer performance in tasks requiring innovative thinking [10]. A meta-analysis on how intrinsic motivation and extrinsic incentives affect performance is provided by [11].

In the past, extrinsic motivations, represented as "carrots and sticks," namely promises of rewards and threats of punishment, were exclusively used to encourage students to learn. These methods were influenced by behaviorist theories, which emphasized the need for external motivation to bring about behavioral changes [12].

However, with the emergence of positive psychology, the focus shifted from extrinsic to intrinsic motivations. According to Bandura, an individual's belief in their perceived ability to achieve specific outcomes (namely, self-efficacy) has a considerable impact on learning and motivation [13]. [14] specifically examined the role of self-efficacy in academic motivation. [15] explored the relationship between growth mindset and intrinsic motivation from a neuroscientific perspective.

According to their Self-Determination Theory (SDT), three key drivers of intrinsic motivation are: 1) Autonomy, the sense of control and agency in one's learning process that instills in learners a sense of ownership; 2) Relatedness, the feeling of being connected to and cared for by others; and 3) Competence, the pursuit of mastery [10]. [16] elaborated on the importance of autonomy, mastery, and purpose for motivation. [17] developed a taxonomy of intrinsic learning motivations consisting of individual intrinsic motivations and interpersonal intrinsic learning motivations.

There have also been many attempts to develop a scale to measure learning motivations reported in the literature. [18] developed a self-report scale to measure intrinsic versus extrinsic motivation in educational settings. [19] constructed and validated the Academic Motivation Scale (AMS) to measure intrinsic, extrinsic, and amotivation in educational contexts. [20] provided new validity evidence for the AMS from the context of language learning. [21] developed the Intrinsic Motivation Inventory (IMI), a multidimensional measurement tool for intrinsic motivation. [22] developed and validated the Motivated Strategies for Learning Questionnaire (MSLQ) that assess intrinsic motivations. [23] developed the Children's Academic Intrinsic Motivation Inventory (CAIMI) to measure intrinsic motivation in elementary and junior high school students. [24] employed psychometric tools to measure intrinsic motivation and its impact on academic outcomes.

However, these efforts have often failed to consider sufficiently broad spectrum of intrinsic learning motivations, rendering them incomplete. This paper aims to: 1) identify a broader spectrum of intrinsic learning motivations in modern-day students, 2) develop a psychometric test based on these motivations, and 3) assess the effectiveness of an educational program using the tool developed. It lays a solid foundation for an education founded on the intrinsic learning motivations of students, which can in turn develop an entrepreneurial mindset in them as shown in [1] and [2].

2 Spectrum of Intrinsic Learning Motivations

We list five intrinsic learning motivations: 1) Joy of knowledge creation, 2) Joy of self-actualization, 3) Joy of self-expression, 4) Joy of playful synergy, and 5) Joy of selfless service in this paper, defining a broad spectrum of intrinsic motivations required entrepreneurial mode of education.

2.1 Joy of Knowledge Creation

The joy of creating new knowledge is apparently hardwired into the human psyche. Children generally derive joy from their inquisitive exploration of the world around them, constantly probing their surroundings with the question, "Why?" Every piece of information they uncover provides them with joy, fueling their curiosity and encouraging further exploration. Every effort to create new ideas, concepts, suggestions, solutions, meanings, applications, recommendations, renderings, interpretations, and more can generate significant intrinsic learning motivation in individuals.

Literature reveals that the joy of discovering new knowledge is an essential element of intrinsic learning motivation. [21] included curiosity-driven exploration and the joy of discovery as essential components of intrinsic motivation. [25] explored the relationship between interest, curiosity, and intrinsic motivation in the learning process. According to [26], curiosity arises when there is a gap between what individuals know and what they want to know, prompting them to actively seek information to fill that gap. [27] outlined how curiosity and interest develop over time.

Education has a universal purpose of transmitting the accumulated wisdom of humanity to the new generation. However, learning the ideas and concepts developed by others may not sufficiently motivate modern-day learners, who live in an age of information with ready access to vast repositories of knowledge. Instead, if learners can use the ideas of others as a springboard for their own creative thinking, such a learning process can generate sufficient intrinsic motivation.

Thus, a redefinition of learning driven by the joy of creation can be developed alongside the effort to gain mastery over the ideas transmitted by previous generation. At any rate, learners must be able to claim ownership of these insights in a way that enhances their intrinsic motivation. The joy of creation is proportional to the level of autonomy learners exercise in the creation process and the extent of ownership they can claim over the generated insight.

2.2 Joy of Self-Actualization

There is a natural progression toward maturation in all living beings as they prepare for reproduction, manifesting as fruition in trees and adulthood in animals, and so on. This process occurs naturally, independent of the conscious efforts, and is facilitated by the supportive ecosystem. However, intellectual, moral, social, or spiritual maturation may require the active participation of the individual for its realization. Individuals are endowed with an intrinsic motivation to grow and develop according to an ideal model, whether it be parents, teachers, or any other role model or icon, as suggested by social learning theories [13].

Abraham Maslow introduced the hierarchy of needs, with self-actualization as the pinnacle of intrinsic motivation [28]. [29] identified achievement as an intrinsic motivation for learning. [30] provided a comprehensive overview of competence and mastery as sources of abundant intrinsic motivation. [31] explored the role of promoting social responsibility in young people, increasing motivation for learning and school achievement.

Working towards a higher purpose in life can generate significant intrinsic learning motivation, even when the task is challenging [16]. According to [32], individuals with substantial goal orientation are more likely to demonstrate deeper intrinsic learning motivations. However, performance goals grounded in the avoidance of failure undermined intrinsic motivation.

Dweck emphasized nurturing a growth mindset in students to enhance intrinsic learning motivation [33]. [34] also underlined the importance of competence as an intrinsic learning motivation, which helps students successfully complete tasks. [16], on the other hand, suggests that intrinsic motivation is sufficient for students to engage in tasks at which they have mastery.

The joy of self-actualization is better expressed as a drive to excel in life, surpassing benchmarks by setting individual records or breaking records set by others. It is a drive to achieve perfection and mastery in skills and abilities, motivating individuals to invest time, energy, and resources. Thus, the joy of self-realization is a powerful intrinsic learning motivation for new-generation learners in the AI age, where they will be required to adapt their mindset, value-set, skillset, and life-set on demand.

2.3 Joy of Self-Expression

Humans have an inherent desire to express their thoughts and communicate themselves and their achievements to others. The joy of self-expression is a uniquely powerful intrinsic learning motivation for new-generation learners, who are born into the age of social media. They want to share moments of discovery and achievement with others through every available medium.

According to constructivist theories of learning, self-expression allows learners to build their own mental models by connecting new information to their existing knowledge and personal experiences. When individuals are given the opportunity to be creative, they become motivated to learn [35]. Additionally, writing about one's thoughts, feelings, and experiences can provide psychological benefits and increase self-awareness [36].

Individuals utilize various means for their authentic self-expression, including art, literature, music, performance, and more. As Howard Gardner points out, individuals choose to express themselves through their unique combination of multiple intelligences [37].

Authentic self-expression requires a state of perfect freedom of mind, leaving the communicator in a state of self-forgetfulness. Such self-forgetful self-expression is native to great poets, artists, performers, and speakers. It attracts others to explore the inner self of the communicator, making it an invitation to communion with the like-minded individuals. They come together and form interest groups, parties, think tanks, round tables, town halls, social media groups, and more. It also connects people with complementary interests, joining hands to accomplish a common project in which they collaborate with natural interest.

In short, creative, authentic, free, and soulful self-expression can serve as a powerful learning motivation for new-generation students.

2.4 Joy of Playful Ssynergy

There is a basic instinct in every individual to be playful, which is especially evident in children. Play liberates individuals from the monotony of daily chores and disengages their minds, creating fertile ground for creativity, often leading individuals into a state of flow and flexibility. This invites incorporation of gamification into education, which fosters playful engagement with the content of learning. Gamification can also act as a tool for simplification of entrepreneurial learning outcomes.

Research has shown that positive emotions play a crucial role in enhancing intrinsic motivation. For example, when students experience joy, fascination, and interest in their learning activities, they are more likely to seek out new challenges and persist in their efforts to achieve academic success [38]. [39] explored the broad impact of play on motivation, learning, and creativity.

According to [40], opportunities for social interaction and collaboration with peers and mentors can enhance intrinsic motivations by creating a supportive and engaging learning environment. [41] investigated the influence of social relationships on student motivation. [42] highlighted the benefits of cooperative learning and social interdependence.

Playfulness also enhances communication, allowing for a free flow of ideas and opening the doors to communion, collaboration, cooperation, and synergy with others. Play is also a way of being together, synergizing through shared activities. Playing with others, being in communion, collaborating, and learning together are powerful intrinsic learning motivations for new-generation learners. Playful synergy contributes to the success of collaborative ventures, which in turn becomes a source of intrinsic motivation, creating a mutually enriching cycle.

In short, playful synergy forms a powerful intrinsic learning motivation for new-generation learners. Play and synergy mutually complement and enhance each other: as play brings forth synergy, synergy fosters more play, and together they generate intrinsic learning motivations.

2.5 Joy of Selfless Service

As social beings, humans possess a quest to belong to a group and contribute to it in their own unique ways. They carry a desire to identify their rightful place in society with distinguishable roles and functions. Even those who prefer to keep aloof from society usually cherish an inner desire to be useful to others. The joy of being significant, wanted, and useful to others is an intrinsic motivation that drives individuals to selfless service.

The identification of one's mission gives individuals a deep sense of authority and authenticity. It enables them to leave a unique mark, their space in history, and a legacy for posterity. In many educational legacies, discovering one's unique purpose in life is considered the key purpose of education. Education is conceived as a process of leading students to a state of enlightenment, where they can become their own source of illumination. It guides them to do what they ought to do, gradually unfolding the deontological purpose of their existence and preparing them for selfless service to others, their calling and mission.

According to [43], when individuals find a particular topic personally relevant and engaging, it enhances their interest and passion for the subject, boosting intrinsic learning motivation. Similarly, connecting learning to their own experiences enhances learners' personal interests and adds meaning to their lives.

[44] identified finding meaning and purpose in life as a powerful motivator. [45] defines an autotelic personality, found in individuals who possess an intrinsic motivation to seek activities for their inherent enjoyment and fulfillment.

In short, learning that contributes to identifying one's rightful place in society and prepares them to occupy it can serve as a powerful intrinsic learning motivation. Similarly, learning that adds meaning to learners' lives generates tremendous intrinsic motivation. The joy of selfless service combines all these aspects and defines itself as a powerful intrinsic learning motivation.

3 Items to Capture Intrinsic Learning Motivations

Based on an extensive literature analysis, we defined a set of intrinsic learning motivations tailored to entrepreneurial education. To measure these intrinsic motivations in learners, we initially drafted eight items for each of the five identified motivations. This initial list was refined through an inter-rater reliability assessment process. In this process, the original statements were distributed among experts and practitioners in education, who evaluated their effectiveness in capturing the intrinsic learning motivations of learners. The final list of statements, determined through inter-rater reliability assessment, is as follows:

3.1 Joy of Creation

The joy of creation is demonstrated through attitudes, values and behaviors relating to creativity, innovation, initiative and quest for autonomous exploration.

1. I put independent effort to develop clarity on a topic, if teachers and textbooks fail to clear my doubts.
2. I put up relentless efforts to explore a question until I get a satisfactory answer.
3. It makes me happy to investigate new aspects of a topic, which are not treated in the classroom.
4. I often find that the discussions taking place in the classroom have already occurred in my mind much prior to it.
5. I like to experiment with new ideas in order to identify or develop newer applications for them.

3.2 Joy of Self-Expression

The joy of self-expression is demonstrated through attitudes, values and behaviors relating to self-expression in multiple ways.

1. I consider myself a creative individual, who can express an idea in a number of ways.
2. I am endowed with necessary skills to communicate my thoughts and views to others with clarity and distinction.
3. I am quick to win the attention of my listeners whenever I start talking to them.
4. I articulate myself well in writing such that I easily win traction with my readers.
5. People used to listen to me for a long time without getting bored, tired or distracted.
6. I usually harvest a good number of 'likes' and 'shares' for my social media posts.
7. I know how to communicate with others using media like dance, poetry, art, and theatre, etc.

3.3 Joy of Self-Actualization

The joy of self-actualization is demonstrated through attitudes, values and behaviors relating to measurable improvement related to one's own talents and potential.

1. Learning keeps empowering me and I am never burdened with my studies.
2. Learning keeps enriching me, and I enjoy every moment of learning without exception.
3. Whatever I learn helps me to grow as a better, confident and competent person.
4. Learning helps me improve myself as a functional individual, rather than just filling the memory cells of my brain.
5. I am careful to develop my skills, values and attitudes based on the insights derived from what I study.
6. My education helps me develop my unique skills in view of my unique vision and mission in life.

3.4 Joy of Playful Synergy

The joy of playful synergy is demonstrated through attitudes, values and behaviors relating to the ability to teamwork for the achievement of common goals.

1. In every discussion with my friends, I participate with joy and eagerness to contribute to it.

2. When I sit with friends to plan and prepare for an event, I find a lot of approvers for my concepts and views.
3. Discussions with my friends help me to improve my own ideas on the topic and to build upon it.
4. I am playful and jovial in my approach to others, such that it is easy for others to work with me.
5. I always develop better solutions to a problem, whenever work with others in a collaborative project.
6. I find greater joy in playing team games rather than solo games.

3.5 Joy of Service

The joy of service is demonstrated through attitudes, values and behaviors relating to one's own contributions to others and society.

1. It makes me proud and happy to join hands with others to do something good for others.
2. I always make an effort to convert the knowledge available with me into something useful for others.
3. I have a clear plan to convert my present studies as an immediate preparation for a unique contribution I would like to make to the welfare of the world at large.
4. I ensure that through my education I equip myself to help myself earn a living and to help others earn their living.
5. I have great respect for all those great people who have contributed to the progress of humankind, and I wish to be one of them.
6. I take care to serve the needs and interests of mother earth in everything I plan and implement during my studies.

Some minor modifications to the wording of the items were also made in consultation with students to enhance their clarity and improve student comprehension.

4 Materials and Methods

4.1 Data Sample

[1] and [2] have developed a pedagogy to cultivate the entrepreneurial mindset of learners, entirely based on the intrinsic learning motivations identified in this paper. The pedagogy described in this book is systematically implemented in the Engineering Physics course at Rajagiri School of Engineering and Technology, India. The questionnaire, developed from the statements listed in the previous section, was administered to students participating in this course at both the beginning of the semester (first measurement point, N = 501; male = 276, female = 225) and the end of the semester (second measurement point, N = 460; male = 254, female = 206). A total of 406 students (male = 219, female = 187) participated in both measurement time points. Therefore, 430 students participated in the questionnaire study before and after a three-month training in Engineering Physics based on the aforementioned pedagogy. Data were also gathered from another group of 30 students who were not introduced to this course.

4.2 Data Analysis

The obtained database was used for exploratory and confirmatory factor analysis and reliability testing to develop a psychometric tool that can assess the effectiveness of training programs and provide insights for further development. Exploratory Factor Analysis (EFA), item characteristics (mean, standard deviation, item difficulty, corrected item-scale correlation), intercorrelations between scales, and internal consistency of scales were performed using the data collected at the first measurement point. Internal consistency was calculated using Cronbach's alpha. The strength of the correlations between each of the five scales was assessed using Pearson correlations.

To evaluate the factorial structure, EFA (principal axis analysis) with orthogonal varimax rotation was performed using parallel analysis, the Scree plot, and the Kaiser-Guttman criterion. The extracted factor structure was then tested through confirmatory factor analysis (CFA) using the data collected at the second measurement point. Established measures and cut-offs were used to assess good model fit: root mean square error of approximation (RMSEA $< .06$), Comparative Fit Index (CFI $> .95$), Tucker-Lewis Index (TLI $> .95$), and Standardized Root Mean Square Residual (SRMR $< .06$) [46]. Differences in model goodness between different models were tested for significance using $\chi 2$-difference tests. Additionally, Akaike's Information Criterion (AIC) and Bayesian Information Criterion (BIC) were used for model comparison, with smaller values indicating a better fit.

Potential changes in intrinsic motivation during the semester as a function of training content were tested for significance using Wilcoxon tests. The Pearson correlation coefficient was also calculated as the effect size in these analyses [47]. All statistical analyses were carried out using R version 4.2.2.

5 Results

5.1 Exploratory Factor Analysis

Using the statistical software R (R Core Team 2022), the developed measurement instrument with $n = 30$ items was subjected to an exploratory factor analysis to investigate the underlying factorial structure and to extract the most important independent factors. The developed questionnaire was therefore administered to students at the beginning of the semester (first measurement point). As part of the data preparation, six subjects were excluded from the analysis due to missing values. The suitability of the sample ($N = 501$, female $= 225$, male $= 276$) for conducting an exploratory factor analysis was confirmed by means of the Kaiser-Meyer-Olkin test (KMO) and Bartlett's test (KMO $= .89$, Bartlett's test: $\chi 2 = 4765.902$, $p < .001$). Examination of the presumed multivariate normal distribution of the data using the Kolmogoriv-Smirnov and Shapiro-Wilk tests shows that all items have a left-skewed distribution instead of a normal distribution. Therefore, a principal axis factor analysis followed by varimax rotation was used as the extraction method for the exploratory factor analysis. This extraction method is recommended when the normal distribution is violated [48]. Using the Kaiser-Gutmann criterion (eigenvalue > 1), the scree plot, and parallel analysis, four factors were finally extracted (43% variance explanation).

A total of 14 items were excluded due to low factor loading ($\lambda < .45$; [49], high secondary loading ($\lambda > .30$), or low commonality ($h2 < .20$) [50]. The selection of individual items was not made solely on the basis of statistical characteristics, but also content-related considerations were considered for scale formation [50]. The final selected items have unique factor loadings between .51 and .72 for one factor. The first and second factors can each be assigned five items, the third factor four items, and the fourth factor three items.

The first factor is assigned items that illustrate the learner's joy in self-actualization and autonomous, lifelong learning (scale: Joy of self-actualization). The items of the second factor reflect the joy and competence of communicating ideas and reflections (scale: Joy of self-expression). The third scale: Joy of playful synergy is assigned items that encompass the joy of playfully engaging with others to further develop one's own ideas. Finally, the Joy of service scale includes items that reflect the joy of developing something useful for others during a learning process and being a valuable group member.

The distribution of the items among four factors thus deviates slightly from the hierarchical structure of the originally developed instrument, which assumes a fifth factor of intrinsic motivation (Joy of knowledge creation).

The results of the factor analysis as well as all relevant (statistical) parameters are summarized in Table 1.

1) Item statistics, internal consistency and correlations

On a descriptive level, it can be seen that relatively high agreement values result for the individual items. The mean values for the four-point (0 "not at all true" - 3 "true exactly") items 1–5 (factor 1: joy of self-actualization) range from 1.65 to 2.17, for items 6–10 (factor 2: joy of self-expression) from 1.59 to 2.00, for items 11–14 (factor 3: joy of playful synergy) from 2.13 to 2.23, and for items 15–17 (Factor 3: joy of service) from 2.24 to 2.38. The standard deviations of the means of the single items (item 1 to item 16) range largely between a half to a whole scale value ($SDMin = 0.54$; $SDMax = .77$). At the scale level, the highest average agreement is shown for the joy of service scale ($M = 2.30$, $SD = .45$), joy of playful synergy ($M = 2.19$, $SD = .45$), followed by the joy of self-actualization scale ($M = 1.88$, $SD = .47$), and finally joy of self-expression ($M = 1.80$, $SD = .49$).

The item difficulty (P%) is additionally considered in order to check whether the items allow for interindividual differences in response behavior. The item difficulty of the individual items is on average 67.1% and varies between 53% and 79.33%. Thus, the item difficulty lies within the mean difficulty range recommended by [51] between 20% and 80%.

The corrected item scale correlation (r) also varies from .39 to .63, indicating moderate to high discriminatory power, thus providing a sufficiently discriminative distinction between individuals with high intrinsic learning motivation and individuals with low intrinsic learning motivation [51]. Table 1 shows the means, standard deviations, item difficulty, and corrected item-scale correlation.

Significant medium-strong correlations also appear between the three scales (joy of self-actualization, joy of self-expression, joy of playful synergy) (Table 2). The joy of self-actualization scale correlates significantly positively with the joy of self-expression

Table 1. Rotated standardized factor matrix (varimax) of the principal axis factor analysis, reliability indices, and descriptive characteristic values of the scales and items (N = 501).

Scale	Item	M (SD)	P%	r	1	2	3	4	h^2
1 Joy of self-actualization	1) Learning helps me improve myself as a functional individual rather than just filling the memory cells of my brain	2.04 (.59)	68	.63	**.72**	-	-		.60
	2) Learning keeps empowering me and I am never burdened with my studies	1.65(.69)	55	.54	**.62**	-	-		.41
	3) Whatever I learn. Helps me to grow as a better. Confident and competent person. And mission in life	2.17(.57)	72.33	.55	**.60**	-	-		.46
	4) Learning keeps enriching me. And I enjoy every moment of learning without exception	1.69(.64)	56.33	.55	**.59**	-	-		.44
	5) My education helps me develop my unique skills in view of my unique vision	2.01(.57)	67	.49	**.52**	-	-		.35
2 Joy of self-expression	6) I articulate myself well in writing such that I easily win traction with my readers	1.63(.68)	54.33	.52	-	**.64**	-		.43

(continued)

Table 1. (*continued*)

Scale	Item	M (SD)	P%	r	1	2	3	4	h^2
	7) I am quick to win the attention of my listeners whenever I start talking to them	1.79(.76)	59.67	.54	-	**.60**	-		.42
	8) I am endowed with necessary skills to communicate my thoughts and views to others with clarity and distinction	2.00(.63)	66.67	. 53	-	**.57**	-		.40
	9) I know how to communicate with others using media like dance. Poetry. Art. And theatre. Etc	1.59(.77)	53	.47	-	**.54**	-		.35
	10) I consider myself a creative individual, who can express an idea in a number of ways	2.00(.65)	66.67	.46	-	**.51**	-		.34
3 Joy of playful synergy	11) In every discussion with my friends. I participate with joy and eagerness to contribute to it	2.19(.58)	73	.55	-	-	**.64**		.49

(*continued*)

Table 1. (*continued*)

Scale	Item	M (SD)	P%	r	1	2	3	4	h^2
	12) I am playful and jovial in my approach with others. Such that it is easy for others to work with me	2.13(.61)	71	.54	-	-	**.60**		.49
	13) Discussions with my friends helps me to improve my own ideas on the topic and to build upon it	2.23(.54)	74.33	.49	-	-	**.53**		.37
	14) I find greater joy in playing team games rather than solo games	2.22(.73)	74	.39	-	-	**.52**		.30
4 Joy of service									
	15) I have great respect for all those great people who have contributed to the progress of humankind, and I wish to be one among them	2.38(.56)	79.33	.57	-	-	-	**.71**	.53

(*continued*)

Table 1. *(continued)*

Scale	Item	M (SD)	P%	r	1	2	3	4	h^2
	16) I take care to serve the needs and interests of the mother earth in everything I plan and implement during my studies	2.24							
(.57)	74.67	.56	-	-	-	**.67**	.51		
	17) Discussions with my friends helps me to improve my own ideas on the topic and to build upon it	2.27(.55)	75.67	.49	-	-	-	**.54**	.38
	SS loadings				2.1	1.9	1.6	1.6	
	Percentage of variance				12	12	9	9	
	Mean value (M)				1.9	1.8	1.8	1.8	
	Standard deviation (SD)				.47	.47	.47	.47	
	Cronbach's Alpha				.78	.74	.70	.72	

Note: *The table presents the mean values (M) and standard deviations (SD) of the items, percentage of agreement (P%), item-factor correlations (r), communalities (h^2), sum of squared (SS) loadings, factor loadings, explained variance, and Cronbach's alpha to assess the internal consistency of the scales*

($r = .340$, $p < .001$) and joy of playful synergy ($r = .315$, $p < .001$) scales. A highly significant positive correlation is also found between the joy of self-expression and joy of playful synergy scales ($r = .369$, $p < .001$) and joy of service and joy of playful synergy scales ($r = .322$, $p < .001$), respectively. A small positive correlation exists between the joy of service and joy of playful synergy scale ($r = .251$, $p < .001$).

The analysis of internal consistency (Cronbach's alpha) confirms that the overall scale (16 items) as well as the three subscales, despite a low number of items (3–5 items), are in an acceptable range with $\alpha < .70$ [52]. Removing one item does not result in higher reliability of the scale. Accordingly, the item analysis as well as the exploratory factor analyses confirm the suitability of the instrument.

Table 2. Scale intercorrelation.

Scale	(1)	(2)	(3)	(4)
(1) Joy of self-actualization	1	-	-	-
(2) Joy of self-expression	.340*	1	-	-
(3) Joy of playful synergy	.315*	.369*	1	
(4) Joy of service	.337*	.251*	.322*	1

* $p < .001$.

5.2 Confirmatory Factor Analysis

For factorial validation, the questionnaire was administered to students again at the end of the semester at a second measurement time point. Based on the results of the exploratory factor analysis at the first measurement time point, the data collected at the second measurement time point (N = 460) were used to conduct a confirmatory factor analysis with R. The latent factor structure was tested in the confirmatory factor analysis (CFA), due to the lack of multivariate normal distribution of the data, by using the robust maximum likelihood estimation (MLR) procedure with Yuan-Bentler correction.

Confirmatory factor analysis of the four-factor model initially reveals fit statistics that remain unsatisfactory ($\chi2[113] = 237.28$, $p < .001$, TLI = .89, CFI = .91, RMSEA = .049, AIC = 12849.6, BIC = 12849.60). Based on identified misspecifications indicated by correlated residuals (error covariances), we gradually incorporated the correlated error terms for the following pairs of content-overlapping items into the model: Items 6 and 8 (EPC = -.08, MI = 16.84), Items 13 and 14 (EPC = .18, MI = 120.09), Items 14 and 15 (EPC = -.06, MI = 18.15), and Items 19 and 21 (EPC = .04, MI = 10.13). The inclusion of these correlations was justified by the conceptual overlap and thematic proximity of the items, which can lead to systematic measurement errors. Specifically, these errors may relate to the transformative impact of learning (Items 1 & 2, 2 & 3), expressive abilities across various communication forms (Items 6 & 8), and the playful, interactive nature of social interactions (Items 11 & 12).

This procedure results in a model with good fit ($\chi2[109] = 126.28$, p >. 05, TLI = .98, CFI = .98, RMSEA = .02, AIC = 12667.86, BIC = 12849.15) that does not reveal any other error specifications. The standardized parameter estimates of the final four-factor model show acceptable high factor loadings (.49 to .77) and were visualized in Fig. 1.

The intercorrelation between the four latent factors is on average r = .377. The joy of self-actualization scale correlates significantly positively with the joy of service (r = .397, $p < .001$), joy of self-expression (r = .348, $p < .001$), and joy of playful synergy (r = .352, $p < .001$) scales. A highly significant positive mean relationship can also be found between the joy of self-expression and joy of playful synergy scales (r = .399, $p < .001$) and joy of service (r = .305, $p < .001$), respectively. Between the scales joy of service and joy of playful synergy a medium strong positive correlation is found (r = .460, $p < .001$).

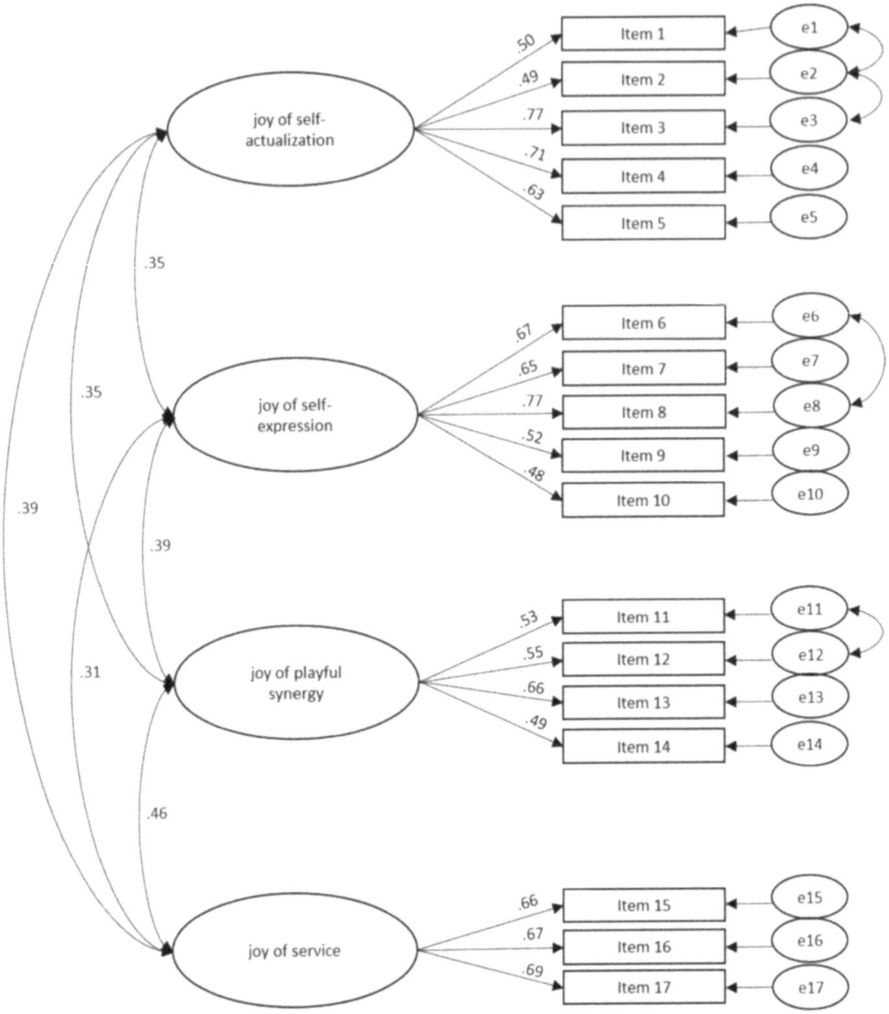

Fig. 1. Results of confirmatory factor analysis for the four-factor model; standardized coefficients, all values $p < .001$.

The internal consistency of the scales according to Cronbach's α is again acceptable for all four scales (joy of self-actualization $\alpha = .78$; joy of self-expression $\alpha = .73$; joy of service $\alpha = .71$; joy of playful synergy $\alpha = .67$). Removal of one item did not result in higher reliability of the scale.

To extend the four-factor model, a single-factor model (general-factor model), a bifactor model, and a higher-order model are established and compared. The general factor model, in which all items load on a single latent dimension, is rejected because significantly worse fit statistics are obtained. The higher order model as well as the bi-factor models (uncorrelated/correlated factors) also show worse model fit. Finally, the model fit indices confirm that the four-factor model fits the data best (Table 3).

Table 3. Fit statistics of the confirmatory factor analyses.

Modell	χ2(df)	CFI	TLI	AIC	BIC	RMSEA	SRMR
General-factor-model	423.81 (115)*	.782	.742	13068.9	13225.5	.077	.078
Four-factor-model	126.28 (109)	.988	.985	12667.8	12849.1	.019	.040
Higher-order-Model	134.01 (111)	.984	.980	12674.5	12847.6	.021	.043
Bi-factor-model (correlated factors)	127.23 (92)*	.975	.963	12701.5	12952.8	.029	.035
Bi-factor-model (uncorrelated factors)	175.97 (102)*	.948	.930	12743.8	12953.9	.040	.048

* p. <. 001

5.3 Training Effectiveness

To test the effectiveness of Entrepreneurial Physics training on intrinsic motivation to learn, students were surveyed at two measurement points (at the beginning and end of the semester) using the developed questionnaire. Students from the Computer Science and Business Systems who did not participate in "Entrepreneurial Physics" course were also surveyed at two measurement time points and served as a control group. To test for training-specific differences with regard to the four identified intrinsic motivational aspects, non-parametric Wilcoxon tests were performed because the normal distribution of the residuals was not met. Consistent with the findings, students with training content ($n = 376$) on Entrepreneurial Physics reported significantly higher Joy of self-actualization scale scores at the end of the semester than at the beginning of the study semester ($V = 24038$, $z = -2.870$, $p = 0.004$, $r = 0.148$). Consistent with expectations, students with training content also report higher scores on the Joy of self-expression scale at the end of the study semester than at the beginning of the study semester ($V = 26727$, $z = -4.448$, $p < 0.001$, $r = .229$). Differences in the other motivational scales Joy of service [$V = 13192$, $z = -0.857$, $p = 0.391$] and Joy of playful synergy [$V = 17184$, $z = - 1.092$, $p = 0.274$] prove not to be statistically significant. In the control group of students with no training content ($n = 30$) over the course of the semester, all motivation scales: Joy of self-actualization [$V = 133$, $z = -1.031$, $p = 0.302$], Joy of self-expression [$V = 85$, $z = -0.385$, $p = 0.701$], Joy of service [$V = 37$, $z = -.562$, $p = 0.574$], Joy of playful synergy [$V = 49.5$, $z = -0.946$, $p = 0.344$] show no statistically significant differences. Boxplot diagrams (Fig. 2) are created for the experimental group (A) and the control group (B) for visual representation of perceived intrinsic learning motivation according to developed questionnaire scale during the semester (two measurement time points).

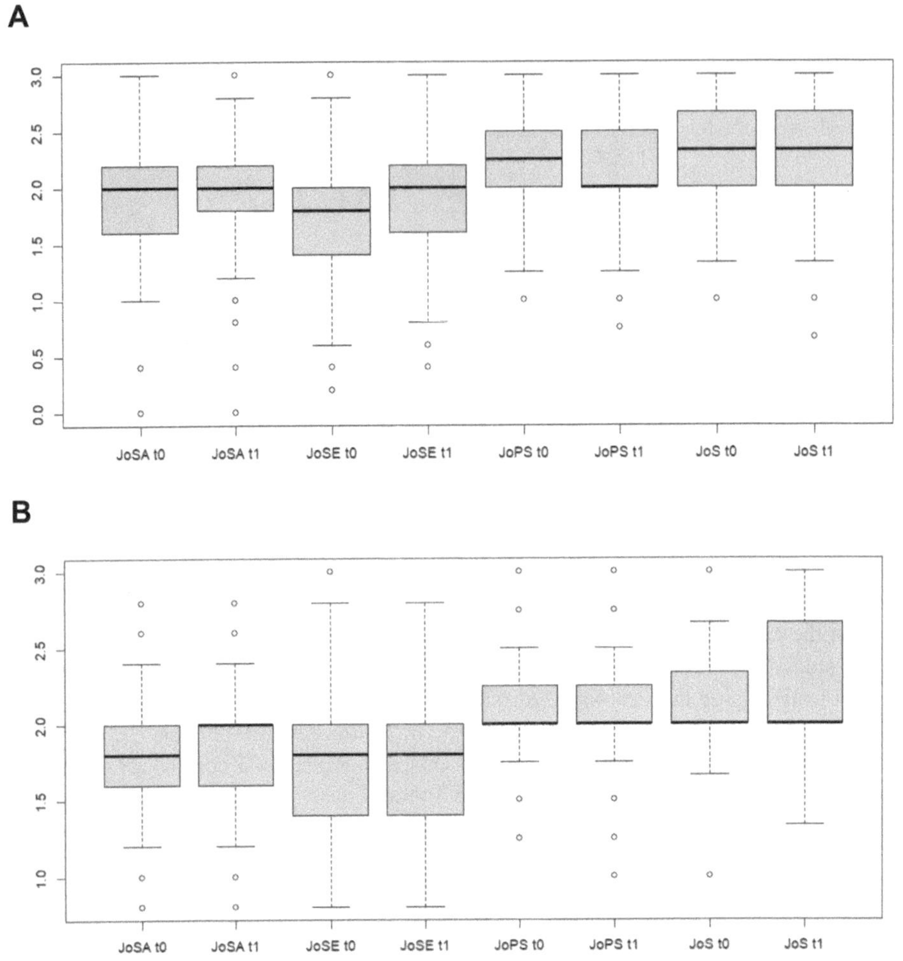

Fig. 2. (A) Boxplots of the scale values (Joy of Self-Actualization [JoSA], Joy of Self-Expression [JoSE], Joy of Playful Synergy [JoPS], Joy of Service [JoS]) for students with training content at two measurement time points (t0, t1). (B) Boxplots of the scale values (Joy of Self-Actualization [JoSA], Joy of Self-Expression [JoSE], Joy of Playful Synergy [JoPS], Joy of Service [JoS]) for students without training content (control group) at two measurement time points (t0, t1).

6 Discussion

The examination of the factorial validity specifically shows that the originally assumed factor structure with the following five factors: 1) Joy of knowledge creation, 2) Joy of self-actualization, 3) Joy of self-expression, 4) Joy of playful synergy, and 5) Joy of selfless service may not hold good entirely. The Exploratory Factor Analysis and the subsequent Confirmatory Factor Analysis suggest that intrinsic motivation to learn can be operationalized using four dimensions (Joy of self-expression, Joy of self-actualization, Joy of playful synergy, Joy of service), each with 3–5 items. The four subscales represent definable constructs. The reliability analysis, which was also carried out, provides satisfactory results and confirms the internal consistency of the instrument's subscales, despite the small number of items.

The joy of knowledge creation by students, widely regarded as a hallmark of higher learning, is being eliminated from the list of intrinsic learning motivations in the current study. However, this does not mean that knowledge production ceases to be an intrinsic learning motivation altogether. Rather, it becomes absorbed into other aspects of intrinsic learning motivations, making it unrecognizable to students. This highlights the need to clearly distinguish the elements associated with the joy of knowledge creation from those describing other intrinsic learning motivations. Additionally, it serves as a critique of the current methods of teaching physics to this group of students, which have failed to enable them to contribute meaningful insights to the discipline. Thus, this study leads to the need for reimagining of physics education—one that empowers students to develop their own insights based on the fundamental laws of physics they learn.

Applying the tool to the students of Entrepreneurial Physics, it is found that the tool is sensitive enough to measure the development of some of the intrinsic learning motivations during the Entrepreneurial Physics program, which lasted for three months only. The results of the assessment can be used to improve the course further and optimize it to generate better results in the development of intrinsic learning motivations in the students.

In summary, the present study makes an important contribution to capturing other relevant aspects of intrinsic motivation that have not been considered in previous research literature. The present study has succeeded in developing an instrument with good test properties for recording relevant aspects of intrinsic motivation. In principle, the questionnaire can be used as a self-assessment instrument to record various components of intrinsic motivation and can also be used for scientific studies that want to investigate and optimize the effectiveness of training courses developed to promote intrinsic motivation.

Nevertheless, limiting aspects of the study must be taken into account. The results are limited to the sample of Indian students and cannot be generalized to other populations and disciplines. Furthermore, as the questionnaire is a self-assessment, systematic bias cannot be completely excluded. Having said that, the validation of the tool is not complete. The convergent validity of the present study remains unclear and should be further examined in follow-up studies. It is also not sure that the apparent improvement of the intrinsic learning motivations in the students will be sustainable.

Since intrinsic motivation to learn is a subjective construct, a self-report questionnaire was used in this case, ensuring that we could do so at minimal expense. We plan to further validate these results using different procedures, such as interviews, creative writing, and case studies. Additionally, we have already implemented student portfolios to support this cross-validation.

7 Conclusion

Education today is understood as more than just the transfer of knowledge. It encompasses not only academic and professional qualifications, but also social, cultural, and personal aspects. The United Nations Educational, Scientific, and Cultural Organization (UNESCO), for example, defines education as a fundamental human right and as the basis for sustainable development, peace, and prosperity. Education is thus understood as a comprehensive process that enables individuals to participate actively and responsibly in society and culture. In the current debate on education, the so-called future skills of the new generation students, the so-called AI-natives are coming to the fore. People need future skills in order to remain capable of acting, successful, and adaptable in an increasingly complex, digital, and dynamic world [53]. These are personal, social, and digital skills that are important across industries and professions and can be applied in a wide range of individual, social, and business contexts [53]. It is obvious that interest and motivation play a central role in education when it comes to acquiring skills.

Education is often understood as the intellectual formation that helps students develop the appropriate mindset and skillset required for the life and workspaces of the future. To assess the development of these mindsets and skillsets, psychometric tests are widely employed. In modern times, education must focus on the development of an entrepreneurial mindset driven by intrinsic learning motivations (Panthalookaran, 2022). A pedagogical approach based on intrinsic learning motivations that generates entrepreneurial learning outcomes can facilitate such mindset development (Panthalookaran, 2023). The current research aimed to develop a psychometric tool for measuring the development of a broad spectrum of intrinsic learning motivations, including 1) Joy of knowledge creation, 2) Joy of self-actualization, 3) Joy of self-expression, 4) Joy of playful synergy, and 5) Joy of selfless service. The psychometric tool was developed based on a systematic analysis of data generated by students participating in an educational program titled Engineering Physics taught based on an entrepreneurial pedagogy. The results of the statistical analysis suggest that a revised tool with fewer items is sufficiently sensitive to capture the development of intrinsic learning motivations in students. However, further research is required to address its limitations and refine it into a robust psychometric instrument capable of accurately measuring all aspects of the variant of intrinsic learning motivation.

References

1. Panthalookaran, V.: Education in a VUCA-driven world: salient features of an entrepreneurial pedagogy. High. Educ. Future **9**(2), 234–249 (2022)
2. Panthalookaran, V.: Pedagogy of the Digital Natives: An Introduction to Entrepreneurial Education. Pallikkutam Publications, Kochi, India (2023)
3. Eccles, J.S., Wigfield, A.: Motivational beliefs, values, and goals. Annu. Rev. Psychol. **53**(1), 109–132 (2002)
4. Froiland, J.M., Worrell, F.C.: Intrinsic motivation, extrinsic motivation, and academic achievement among adolescents. J. Educ. Res. **109**(6), 671–682 (2016)
5. Schunk, D.H., Pintrich, P.R., J.L., Meece, J.L.: Motivation in education: theory, research, and applications. 3rd ed. Upper Saddle River, NJ: Pearson/Merrill Prentice Hall (2008)
6. Stipek, D.J.: Motivation and instruction. Handbook Educ. Psychol. **1**, 85–113 (1996)
7. Ryan, R.M., Deci, E.L.: Intrinsic and extrinsic motivations: classic definitions and new directions. Contemp. Educ. Psychol. **25**(1), 54–67 (2000)
8. Deci, E.L.: Effects of externally mediated rewards on intrinsic motivation. J. Pers. Soc. Psychol. **18**(1), 105–115 (1971)
9. Deci, E.L., Koestner, R., Ryan, R.M.: A meta-analytic review of experiments examining the effects of extrinsic rewards on intrinsic motivation. Psychol. Bull. **125**(6), 627–668 (1999)

10. Ryan, R.M., Deci, E.L.: Self-Determination Theory: Basic Psychological Needs in Motivation, Development, and Wellness. Guilford Publishing, New York (2017)
11. Cerasoli, C.P., Nicklin, J.M., Ford, M.T.: Intrinsic motivation and extrinsic incentives jointly predict performance: a 40-year meta-analysis. Psychol. Bull. **140**(4), 980–1008 (2014)
12. Pavlov, I.P.: Conditioned reflexes: an investigation of the physiological activity of the cerebral cortex. Reprint of 1927 ed., Martino Fine Books (2015)
13. Bandura, A.: Social Learning Theory. 1st ed. Pearson (1977)
14. Schunk, D.H.: Self-efficacy and academic motivation. Educ. Psychol. **26**(3–4), 207–231 (1991)
15. Ng, B.: The neuroscience of growth mindset and intrinsic motivation. Brain Sci. **8**(2), 20 (2018)
16. Pink, D.H.: Drive: The Surprising Truth About What Motivates Us. Penguin (2011)
17. Malone, T.W., Lepper, M.R.: Making learning fun: a taxonomy of intrinsic motivations for learning. In: Aptitude, Learning, and Instruction, Routledge, pp. 223–254 (2021)
18. Harter, S.: A new self-report scale of intrinsic versus extrinsic orientation in the classroom: motivational and informational components. Dev. Psychol. **17**(3), 300–312 (1981)
19. Vallerand, R.J., et al.: The academic motivation scale: a measure of intrinsic, extrinsic, and amotivation in education. Educ. Psychol. Meas. **52**(4), 1003–1017 (1992)
20. Tseng, W.T., Liu, H., Nix, J.: Validation of the revised Academic Motivation Scale (AMS) in the context of language learning. Psychol. Rep. **120**(2), 394–414 (2017)
21. Deci, E.L., Ryan, R.M.: Intrinsic Motivation and Self-determination in Human Behavior. Plenum, New York, NY (1985)
22. Pintrich, P.R., Smith, D.A.F., Garcia, T., McKeachie, W.J.: Reliability and predictive validity of the Motivated Strategies for Learning Questionnaire (MSLQ). Educ. Psychol. Meas. **53**(3), 801–813 (1993)
23. Gottfried, A.E.: Academic intrinsic motivation in elementary and junior high school students. J. Educ. Psychol. **77**(6), 631–645 (1985)
24. Froiland, J.M., Oros, E.: Intrinsic motivation, perceived competence and classroom engagement as longitudinal predictors of adolescent reading achievement. Educ. Psychol. **34**(2), 119–132 (2014)
25. Schiefele, U.: Interest, learning, and motivation. Educ. Psychol. **26**(3–4), 299–323 (1991)
26. Loewenstein, G.: The psychology of curiosity: a review and reinterpretation. Psychol. Bull. **116**(1), 75–98 (1994)
27. Hidi, S., Renninger, K.A.: The four-phase model of interest development. Educ. Psychol. **41**(2), 111–127 (2006)
28. Maslow, A.H.: Motivation and Personality. New York, NY, Harper (1954)
29. Nicholls, J.G.: Achievement motivation: conceptions of ability, subjective experience, task choice, and performance. Psychol. Rev. **91**(3), 328–346 (1984)
30. Elliot, A.J., Dweck, C.S., eds.: Handbook of Competence and Motivation. Guilford Publications (2013)
31. Ford, M.E.: Motivating Humans: Goals, Emotions, and Personal Agency Beliefs. Sage Publications, Newbury Park, CA (1992)
32. Elliot, A.J., Harackiewicz, J.M.: Approach and avoidance achievement goals and intrinsic motivation: a mediational analysis. J. Pers. Soc. Psychol. **70**(3), 461–475 (1996)
33. Dweck, C.S.: Mindset: The New Psychology of Success. Random House, New York, NY (2006)
34. Ames, C., Archer, J.: Achievement goals in the classroom: students' learning strategies and motivation processes. J. Educ. Psychol. **80**(3), 260–267 (1988)
35. Fasko, D.: Education and creativity. Creativity Res. J. **13**(3–4), 317–327 (2001)
36. Pennebaker, J.W., Chung, C.K.: Expressive Writing: Connections to Physical and Mental Health. Oxford Univ. Press (2011)

37. Gardner, H.E.: Frames of mind: the theory of multiple intelligences. Basic Books (2011)
38. Fredrickson, B.L.: The role of positive emotions in positive psychology: the broaden-and-build theory of positive emotions. Am. Psychol. **56**(3), 218–226 (2001)
39. Brown, S., Vaughan, C.: Play: how it shapes the brain, opens the imagination, and invigorates the soul. New York, NY, Avery (2009)
40. L. S. Vygotsky, Mind in Society: The Development of Higher Psychological Processes, vol. 86. Harvard Univ. Press, 1978
41. Wentzel, K.R.: Social-motivational processes and interpersonal relationships: implications for understanding motivation at school. J. Educ. Psychol. **91**(1), 76–97 (1999)
42. Johnson, D.W., Johnson, R.T.: An educational psychology success story: social interdependence theory and cooperative learning. Educ. Res. **38**(5), 365–379 (2009)
43. Renninger, K.A., Hidi, S.: The Power of Interest for Motivation and Engagement. Routledge, New York, NY (2016)
44. Frankl, V.E.: Man's Search for Meaning: An Introduction to Logotherapy. Beacon Press, Boston, MA (1963)
45. Csikszentmihalyi, M.: Finding Flow: The Psychology of Engagement with Everyday Life. Basic Books, New York, NY (1997)
46. Hu, L.T., Bentler, P.M.: Cutoff criteria for fit indexes in covariance structure analysis: conventional criteria versus new alternatives. Struct. Equ. Model. **6**(1), 1–55 (1999)
47. Cohen, J.: Statistical Power Analysis for the Behavioral Sciences. Taylor and Francis, Hoboken (1988)
48. Costello, A.B., Osborne, J.W.: Best practices in exploratory factor analysis: four recommendations for getting the most from your analysis. Pract. Assess. Res. Eval. **10**(7), 1–9 (2005)
49. Tabachnick, B.G., Fidell, L.S.: Using Multivariate Statistics. Pearson (2007)
50. Bühner, M.: Einführung in die Test-und Fragebogenkonstruktion. Pearson (2011)
51. Bortz, J., Döring, N.: Forschungsmethoden und evaluation: Für Human- und Sozialwissenschaftler. Springer-Medizin-Verlag (2006)
52. Mohamad, M.M., Sulaiman, N.L., Sern, L.C., Salleh, K.M.: Measuring the validity and reliability of research instruments. Procedia-Soc. Behav. Sci. **204**, 164–171 (2015)
53. OECD [Organisation for Economic Co-operation and Development] Future of Education and Skills 2030/2040 (2025). https://www.oecd.org/en/about/projects/future-of-education-and-skills-2030.html

Integration of Artificial Intelligence into mathematics Education at the Middle and Secondary Levels: Obstacles, Opportunities, and Prospects

Abdoulaye Wakhab Diop[1(✉)], Ndeye Massata Ndiaye[1], and Mouhamadou Lamine Ba[2]

[1] Université Numérique Cheikh Hamidou Kane, Dakar, Sénégal
{Abdoulayewakhab.diop,Ndeyemassata.ndiaye}@unchk.edu.sn
[2] Ecole Supérieur Polytechnique de Dakar, Dakar, Sénégal
Mouhamadoulamine.ba@esp.sn

Abstract. Mathematics education in Senegal faces persistent challenges: poor results, growing disinterest in science subjects, and regional inequalities. In this context, the integration of artificial intelligence (AI) into mathematics education at the intermediate and secondary levels is a rapidly growing field of research with significant implications for the education sector. This article offers a forward-looking perspective on the use of AI in mathematics education in Senegal.

A review of the literature reveals that AI offers promising opportunities to improve mathematics learning, particularly by providing adaptive feedback, personalized learning, and interactive learning tools. Examples of existing applications such as Khan Academy and ALEKS illustrate the potential of AI to create stimulating learning environments tailored to the individual needs of students.

However, the use of AI in mathematics education also raises important ethical questions, particularly about transparency, algorithmic bias, privacy, and fairness. By taking a balanced and thoughtful approach, it is possible to fully leverage the benefits of AI while ensuring its ethical and responsible use in middle and high school mathematics education. The results show that if AI is integrated in a way that takes local context into account and is accompanied by informed governance, it could enhance personalization, facilitate remediation, increase motivation, and improve academic achievement.

Keywords: Artificial intelligence · mathematics education · middle schools · high schools · personalized learning · ethical issues

1 Introduction

In Senegal, mathematics is one of the most feared and difficult subjects in middle and high school. Each year, the results of national exams such as the "Brevet de Fin d'Etudes Moyennes (BFEM)" or the baccalaureate reveal a worrying and concerning reality: success rates in mathematics remain low, disinterest in scientific subjects is growing, and the predominance of literary streams is confirmed.

M. E. Auer et al. (Eds.): IMCL 2025, LNNS 1954, pp. 111–121, 2026.
https://doi.org/10.1007/978-3-032-24998-2_9

In 2023, according to official data from the Ministry of Education, 76% of high school graduates came from literary streams, compared to only 24% from scientific and technical streams. This imbalance has serious consequences for the country's economic and technological development. It hinders the development of technical skills, limits local innovation, and increases the mismatch between the training provided and the real needs of the labor market.

Among the factors explaining this disproportionate trend, the teaching of mathematics plays a central role. A fundamental discipline for scientific studies, mathematics is often perceived by students as abstract and elitist. It represents more of an obstacle than a springboard. Its teaching, still marked by traditional, uniform, and transmission-centered pedagogical approaches, contributes to a growing disinterest in scientific fields. These difficulties highlight the urgent need to implement innovative teaching approaches capable of responding to the diversity of students' learning needs and supporting teachers in their practices.

Faced with this worrying trend, it is imperative to rethink the teaching of mathematics in Senegal to make it more inclusive, more motivating, and better suited to the varied profiles of learners. In this context that artificial intelligence is attracting growing interest. The study is based on the hypothesis that integrating AI into mathematics education can help reduce failure rates while strengthening student engagement and success in mathematics in Senegal.

Recent advances in AI suggest profound pedagogical transformations, particularly through personalized, adaptive, and interactive learning tools (Mahon, using virtual worlds to enhance learning and teaching, 2010). However, the enthusiasm generated by these innovations should not obscure the many challenges they pose. Some researchers, such as (Barth, 2013), point out that the deep cognitive processes involved in learning abstract mathematics remain difficult to stimulate through algorithms. Others highlight that integrating AI into education requires strong change management and educational leadership skills (Fraser, 2016). Furthermore, to ensure equitable and sustainable implementation, it is essential to consider issues of school governance (Chan, 2012), the inclusion of students with disabilities (Marquis, 2012), and the diversity of learning profiles.

This article has three main objectives: to take stock of the Senegalese education system in terms of mathematics teaching; to analyze the possibilities for integrating AI in this context; and to identify the obstacles, opportunities, and prospects for appropriate and sustainable implementation in Senegal.

2 Literature Review

The academic literature on the use of artificial intelligence (AI) in education has made significant progress over the past decade. It covers technological, pedagogical, ethical, and political aspects. This section presents a summary of the main international and African contributions, with a focus on mathematics education.

In developed countries, AI is gradually being integrated into digital learning environments. In the United States, for example, platforms such as Assessment and Learning in Knowledge Spaces (ALEKS) and DreamBox Learning have proven effective in improving student performance in mathematics through adaptive algorithms capable of personalizing learning paths (Johnson, 2017).

In Europe, approaches are more cautious, particularly due to ethical concerns related to data protection. Pilot projects supported by the European Union are exploring the use of AI in personalizing learning and formative assessment. However, the emphasis is on compliance with the GDPR and the need to ensure the fairness and transparency of algorithms (Lee, 2020).

On a theoretical level, several models support the integration of AI into education:

- Adaptive learning is based on the ability of systems to adjust the content and difficulty of exercises in real time according to the student's profile (Smith A. B., 2019).
- The content and difficulty of exercises in real time based on the student's profile (Smith A. B., 2019)
- Learning Analytics use the digital traces generated by learners
- To analyze their behavior and predict their future performance (Li, 2021).
- Interactive simulations and immersive environments promote better
- Understanding of abstract concepts in mathematics and science (Zhang, 2016).

All these modules are relevant and make teaching mathematics more accessible.
In Africa, the use of AI in education is still limited but is rapidly emerging.

- in Morocco, several initiatives are exploring the use of smart digital tools to strengthen science and mathematics education in pilot high schools
- in Rwanda, the government has launched an ambitious strategy to integrate AI into various sectors, including education, through partnerships with international EdTech companies
- in South Africa, university projects are using learning analytics to analyze student learning behaviors and identify early risks of dropout

All these experiences show that, despite numerous constraints, there is a growing desire on the African continent to explore the potential of AI in education.

Although the international literature is extensive, it remains dominated by North American, European, and Asian contexts. Few studies focus on West Africa, and even fewer on Senegal. Most of the existing research concentrates on technological aspects, often neglecting the educational, cultural, and social dimensions specific to developing countries.

In the case of Senegal, several issues remain understudied:

- how can AI solutions develop in Western contexts be adapted to an environment characterized by limited resources and a high degree of student heterogeneity?
- what strategies should be put in place to train teachers in the use of these tools?
- how can we ensure inclusive integration that does not reinforce inequalities between urban and rural areas?

All these gaps justify the relevance of a study focused on Senegal, which could serve as a reference case for other African countries facing similar challenges.

3 Statement of the Problem: Challenges in Mathematics Education in Senegal

In Senegal, teaching mathematics at the middle and secondary levels is one of the greatest challenges facing the education system. It faces a series of structural, pedagogical, and cultural challenges. Although considered an essential subject for the country's scientific, technological, and economic development, it remains, paradoxically, one of the most feared and least successful subjects among students. This situation is a cause for concern for public decision-makers, teachers, and education researchers alike, as it reflects deep-rooted structural, pedagogical, and cultural issues.

The results of national and international assessments show a low level of mathematics achievement. According to statistics from the Ministry of National Education (Senegal, 2022), approximately 60% of students fail the BFEM mathematics test, and the average pass rate for mathematics in the baccalaureate is no higher than 20.5%. These figures, which have remained stable for several years, reveal a structural inability to significantly improve performance.

In addition, the imbalance between the literary and scientific streams is growing. In 2023, nearly 76% of baccalaureate candidates came from the arts streams, compared to only 24% from the science and technology streams. This massive disinterest in science streams not only compromises the development of STEM skills but also limits young Senegalese people's access to emerging professions related to digital technology, engineering, and applied sciences.

In terms of teaching methods, mathematics instruction still relies heavily on transmissive methods, where the teacher possesses the knowledge and the student adopts a passive stance. This approach, inherited from the colonial system and ill-suited to contemporary needs, emphasizes memorization and repetitive exercises at the expense of conceptual understanding and practical application.

Teachers often express a lack of continuing education and teaching support. The use of digital tools, although promoted by certain reforms, remains marginal and unevenly distributed. Individualized remediation is rarely practiced due to the lack of time and resources, even though it is important for students with difficulties.

Structural challenges exacerbate the situation that Senegal faces:

- a shortage of qualified teachers: Mathematics classes, particularly in rural areas, are often taught by contract or unqualified teachers
- overcrowded classrooms: Some classes have more than 70 students, making individualized support impossible
- a marked digital divide: many rural areas remain on the margins of the digital revolution
- insufficient teaching resources: lack of textbooks, absence of a math lab, scarcity of interactive tools

All these constraints reduce the education system's ability to adapt to new educational and technological needs.

Beyond the technical aspects, the perception of mathematics in the Senegalese society plays a crucial role. The subject is often associated with difficulty, failure, and elitism. Students, influenced by those around them, sometimes develop an early aversion to mathematics, which they see as an almost insurmountable obstacle to their academic success.

This negative perception is reinforced by the lack of social recognition for scientific careers. While literary professions (law, social sciences, and communication) are perceived as more accessible and prestigious, scientific professions suffer from a lack of image and attractiveness.

Compared to other African countries, Senegal lags in promoting scientific fields of study. For example, in Rwanda, nearly 45% of upper secondary school students pursue scientific fields, thanks to a proactive policy of promoting STEM. In South Africa, public-private initiatives have strengthened mathematics education by using digital tools and targeted academic support programs.

This comparison highlights the urgent need for Senegal to rethink its education policies to align its practices with the economic and technological development ambitions expressed in strategic plans such as the Emerging Senegal Plan (PSE) and the Senegal Digital 2025 plan.

4 Proposed Methodology

The methodology adopted in this research is based on a mixed approach, combining documentary analysis, comparative evaluation of artificial intelligence platforms applied to mathematics education, and a comparison with the Senegalese context. This study aims to identify concrete avenues for integrating AI in middle and secondary education.

4.1 Research Approach

The first step consists of a systematic literature review of scientific work, institutional reports, and pilot projects on the integration of AI into education. Academic databases (Springer, IEEE, Elsevier) as well as reports from the Senegalese Ministry of Education and international organizations (UNESCO, World Bank, OECD) served as the main sources.

The second stage is based on a comparative analysis of five AI-integrated mathematics teaching platforms:

- Khan Academy (free, personalized mathematics module),
- DreamBox Learning (adaptive platform widely used in the United States),
- ALEKS, intelligent learning assessment system
- Photomath (mobile application for interactive problem solving and explanation)
- GeoGebra (mathematical visualization and simulation software)

Finally, a critical comparison was made between the results of this analysis and the realities of the Senegalese education system, considering the infrastructure, teacher training levels, and national priorities in terms of digital education.

4.2 Platform Evaluation Criteria

To ensure the relevance and rigor of the analysis, five main criteria were selected:

- personalization of learning: the platform's ability to adapt content and pace to the specific needs of each student
- accessibility and inclusivity: cost, offline availability, and compatibility with different devices (computers, tablets, smartphones)
- interactivity and motivation: presence of simulations, fun elements, and real-time feedback
- pedagogical relevance: alignment with official mathematics curricula and ability to develop reasoning skills

All these criteria, inspired by international literature, allow for a balanced comparison between solutions developed in different contexts but potentially transferable to Senegal.

4.3 Comparative Analysis Process

Each platform was studied based on the following:

- scientific journals (peer-reviewed articles)
- evaluation reports produced by ministries or educational bodies in the user countries
- practical tests carried out by teachers and learners during documented pilot projects

A comparative table summarizing the strengths and limitations of each solution was developed, making it possible to identify the applications best suited to the Senegalese context.

4.4 Limitations

Like any other research, this study has several limitations:

- lack of field testing: In the absence of effective implementation in Senegalese institutions, the analysis remains based on secondary data
- variability of contexts: The platforms studied were developed for different educational environments, which limits their direct transferability
- incomplete data: the available national statistics do not always allow for a detailed assessment of mathematical skills, which complicates the correlation with the proposed solutions
- unresolved ethical issues: Most platforms do not communicate transparently about personal data management, an important aspect for their adoption in Senegal

4.5 Justification of the Methodological Choice

Despite these limitations, the methodology adopted has several positive points:

- putting international practices into perspective with local needs
- identify the minimum conditions required for the successful adoption of AI in mathematics
- providing a solid basis for future pilot experiments in Senegalese schools

This methodology is an important step in building a reference framework for integrating AI into mathematics education in Senegal.

5 Results

Application	Strengths	Weaknesses	Use of AI
ALEKS	• Adaptive learning that personalizes according to students needs • Intuitive and easy-to-use interface • Wide coverage of math topics	• High subscription cost-Lack of student-to-student interaction • Limited interaction for students interested in technology	• Adaptive assessment : Uses AI to evaluate students knowledge and adapt questions accordingly • Personalization : Creates personalized learning paths based on students progress and skills
Khan Academy	• Free and accessible to everyone • High-quality educational videos • Interactif exercises and progress tracking	• Requires internet connection • Less personalization compared to other apps • Limited professional supervision of students	• Content recommendation : Use AI to suggest videos and exercises according to student progress • Data Analysis : Analyzes student responses to identify weak areas and measure difficulties
Photomath	• Allow solving math problems using the phone's camera-Provides detailed explanation • Simple, intuitive, attractive interface	• Less effective for complex problems requiring deep understanding • May encourage dependence on automated solutions • Limited usefulness beyond high school math	• Text recognition : Use OCR (Optical character recognition) to read handwritten or printed math problems from photos • Automatic resolution : AI solves problems and provides step-by-step explanations
DreamBox Learning	• Highly interactive and engaging for students • Adaptive, adjusting learning pace for each student • available in multiples languages	• High cost for schools and families • Requires strong internet connection • Learning experience limited without optimal connectivity	• **Adaptative learning** : AI customizes teaching content in real time based on student needs and performance • **Predictive analysis** : AI analyzes learning patterns to anticipate difficulties and adjust lessons accordingly

The analysis shows that.

- Khan Academy and Photomath stand out for their accessibility and free access, which makes them particularly interesting for expanding access to educational resources
- DreamBox and ALEKS offer highly advanced personalization, but their cost and dependence on the digital infrastructure limit their adoption
- GeoGebra is an intermediate solution that is free but requires educational support

In the Senegalese context, three main observations emerge:

- accessibility and equity: Free tools such as Khan Academy, Photomath, and GeoGebra are the most suitable, if access to digital equipment and connectivity can be guaranteed
- teacher training: The effectiveness of these tools depends largely on the teachers' ability to integrate them into their teaching practices. However, continuing education in digital education remains limited

- governance and ethics: The adoption of AI solutions in education must be governed by public policies that guarantee data protection, transparency, and alignment with national curricula

Beyond the pedagogical aspects, the analysis highlights several issues, such as the need to ensure the protection of students' personal data, the risk of widening the digital divide between urban and rural schools, and dependence on foreign tools that could hinder the development of local solutions.

All these findings confirm that the integration of AI into mathematics education in Senegal is both promising and complex, requiring a gradual and contextualized approach.

6 Discussion

The results of the comparative analysis highlight both the transformative potential of artificial intelligence (AI) in mathematics education and the complex challenges of integrating it into the Senegalese context. This discussion offers a critical interpretation of these results in four areas: opportunities, conditions for success, associated risks, and strategic recommendations.

For Senegal, the integration of AI into mathematics education opens several positive prospects:

- improved learning quality: With adaptive systems and immediate feedback, AI makes it possible to better target students' difficulties and provide personalized solutions.
- reduction of inequalities: Free tools such as Khan Academy and GeoGebra can democratize access to high-quality educational resources, particularly in rural areas.
- increased student motivation: Interactive and fun approaches offered by certain platforms encourage engagement and reduce anxiety about mathematics.
- support for teachers: AI can lighten certain repetitive tasks and allow teachers to focus on providing educational support.
- alignment with national ambitions: The Senegal Digital Plan 2025 and the National Strategy for Education and Training (SNEF) emphasize the importance of digital technology in education, which could create a political environment conducive to innovation.

For the successful integration of AI into mathematics education, several conditions must be considered:

- digital infrastructure: Guaranteeing stable access to the Internet and digital equipment is essential.
- teacher training: Success depends on teachers' ability to effectively integrate these tools into their teaching practices. Initial and continuing training programs must be implemented.
- curriculum alignment: AI solutions must be adapted to Senegalese curricula and incorporate local cultural and linguistic specificity.
- clear regulatory framework: A national policy governing the use of AI in education must be developed, including guidelines on personal data protection and algorithm transparency.

- multi-stakeholder partnerships: The involvement of ministries, universities, EdTech start-ups, and international partners is essential to pool resources and expertise.

Several risks must be considered:

- increased digital divide: if only certain schools have the necessary resources, AI risks worsening inequalities rather than reducing them.
- technological dependence: Excessive reliance on foreign solutions could limit the development of local initiatives and reduce the country's educational sovereignty.
- algorithmic bias: Platforms developed in other cultural contexts may reproduce biases, leading to implicit discrimination.
- risk of dehumanization of teaching: AI cannot replace the role of mediator and guide played by the teacher.
- financial sustainability: paid solutions (DreamBox, ALEKS) represent a high cost for the state or families.

The successful integration of artificial intelligence into education cannot be achieved in isolation; it requires the establishment partnerships. Ministries, universities, EdTech start-ups, and international partners represent complementary actors whose combined resources and expertise are indispensable for fostering systemic innovation. Such collaborations create the enabling conditions for both large-scale experimentation and sustainable policy implementation. Several strategic recommendations can be derived from this observation. First, the development of a national framework for AI in education is essential to provide coherence, establish regulatory safeguards, and ensure alignment with broader socio-economic priorities. Second, the training of teachers and the systematic reinforcement of their digital competences constitute a critical prerequisite, as educators play a pivotal role in mediating technological adoption and shaping learner's experiences. Third, priority should be given to the design of contextually appropriate local solutions, capable of addressing specific cultural, pedagogical, and infrastructural realities highlights the need for targeted policies to reduce the digital divide, including disparities in connectivity, access to devices, and digital literacy. Finally, the governance of AI in education must be grounded in ethical and inclusive principles, ensuring transparency, accountability, and the equitable distribution of benefits all segments of society.

This research will contribute in two ways:

- scientific: It offers a contextualized analysis of AI applied to mathematics education in an African country, filling a gap in the literature dominated by Western and Asian contexts.
- practical and political: It provides concrete recommendations for educational decision-makers, highlighting the conditions for success and the risks associated with the integration of AI.

7 Conclusion

Mathematics education in Senegal is at a turning point. Despite its strategic importance for the development of scientific and technological skills, the discipline is currently facing poor results, a marked lack of interest among students, and persistent structural constraints. Despite these challenges, artificial intelligence (AI) appears to be a major opportunity to redefine teaching practices and strengthen the effectiveness of the education system.

The results show that AI can offer concrete solutions for improving mathematics learning. Platforms such as Khan Academy and GeoGebra facilitate access to interactive and free resources, while more sophisticated systems such as DreamBox and ALEKS demonstrate the effectiveness of adaptive learning and automated diagnosis. However, their adoption in Senegal faces major obstacles: high cost, the need for a robust digital infrastructure, lack of teacher training, and the absence of a clear regulatory framework.

Thus, AI cannot be seen as a miracle solution. It must be integrated gradually, thoughtfully, and in a manner adapted to the local context. The success of such integration depends on several essential conditions:

- Investment in digital infrastructure to reduce the divide between urban and rural areas
- Continuing education for teachers to equip them with the skills necessary to use intelligent tools for teaching
- Aligning AI solutions with national curricula to ensure their educational relevance
- The development of an ethical and regulatory framework to ensure the protection of student data and the transparency of algorithms

Looking ahead, several avenues of research are worth exploring:

- The implementation of pilot experiments to assess the impact of AI on student achievement in mathematics in Senegal
- The study of the sociocultural aspects of the integration of educational technologies, in particular the perceptions of students and teachers
- Development of local and multilingual AI solutions, integrating national languages to enhance inclusivity
- Assessing the ethical and psychological impact of the use of AI in learning

Ultimately, the integration of AI into mathematics education in Senegal should not be seen as an end, but as a lever for a broader educational project aimed at improving the quality, equity, and relevance of education. By combining technological innovation, political commitment, and the involvement of educational stakeholders, Senegal can become an African model for the responsible and inclusive use of AI in education.

References

Barth, B.: Knowledge in the making, 2nd edn. Retz, Paris (2013)

Chan, T.R.: School leadership and administration. Waveland Press, Long Grove, IL (2012)

Chen, X.: Teacher digital competencies for integrating AI in mathematics teaching. J. Educ. Technol. Dev. Exchange **14**(1), 55–70 (2021)

Chen, L.W.: Early warning systems in education: predicting student performance using machine learning. Int. J. Artif. Intell. Educ. **28**(3), 253–274 (2018)

Fraser, R.S.: Instructional designers' perceptions of the future of learning. Can. J. Learn. Technol. **42**(2), 1–18 (2016)

Johnson, T.: Adaptive learning systems and individual differences: a review. J. Educ. Psychol. Technol. **23**(4), 112–128 (2017)

Jones, M.W.: Using AI to assess students' learning paths in mathematics. Comput. Educ. **148**, 103802 (2020)

Lee, H.: Ethical implications of AI in education: data privacy, bias, and fairness. AI Soc. **35**(3), 645–658 (2020)
Li, S.K.: Predictive analytics in secondary education. Proceedings of the Educational Data Mining Conference, pp. 145–154 (2021)
Mahon, F.C.: Using virtual worlds to enhance learning and teaching. Br. J. Edu. Technol. **41**(1), 10–20 (2010)
Marquis, E.G.: Toward a model for integrating students with disabilities at university. Can. J. Educ. **35**(2), 164–192 (2012)
Sénégal, M.d.: Report on the results of the 2022 baccalaureate City, Dakar. MEN (2022)
Oishi, L.S.: Title: scenario-based collaborative quizzes. IEEE Trans. Educ. **60**(2), 123–130 (2017)
European Union, U.: General Data Protection Regulation (EU 2016/679). European Union, Brussels (2018)
Smith, R.: Instructional design for diverse learners: strategies for inclusive mathematics education. J. Math. Educ. **49**(1), 45–62 (2018)
Smith, A.B.: Personalized learning through artificial intelligence. J. Learn. Anal. AI Educ. **7**(2), 75–90 (2019)
Wang, J.: Integrating AI tools in mathematics instruction. Educ. Tech. Res. Dev. **67**(4), 895–913 (2019)
Werner, H.: Artificial intelligence in education: promises and implications. J. Educ. Technol. Syst. **38**(4), 415–430 (2010)
Zhang, X.H.: Enhancing conceptual understanding with interactive simulations. Interact. Learn. Environ. **24**(6), 1225–1240 (2016)

A Benchmark Study of Machine Learning Models for Predicting Cardiovascular Risks

Adela-Antonia Dineti[1], Bianca-Andreea Balea[1], Ioana Corina Bogdan[1](✉), Horia Alexandru Modran[1], and Doru Ursutiu[1,2]

[1] Faculty of Electrical Engineering and Computers Science, Department of Electronics and Computers, Transilvania University of Brasov, Brasov, Romania, Brasov, Romania
{adela.dineti,andreea.balea}@student.unitbv.ro, {corina.bogdan, horia.modran,ursutiu}@unitbv.ro

[2] Academy of Romanian Scientists AOSR, Str. Ilfov 3, Bucharest, Romania

Abstract. Given the rising incidence of cardiovascular illnesses and related risk factors, the development of precise and effective prediction systems is of great interest, thanks to advancements in artificial intelligence and smart devices. This work evaluates the fulfillment of machine learning algorithms, such as Logistic Regression, Support Vector Machine, K-Nearest Neighbours, Multilayer Perceptron, Random Forest, and XGBoost, for determining the presence of possible heart disease using a publicly available clinical dataset related to cardiovascular disease and risks. For machine learning algorithms (MLA) comparison, precision, recall, F1-score, and overall accuracy metrics were included, and a Pearson correlation matrix was computed to establish a linear relationship among the 14 clinical features, excluding the target variable, in the updated heart disease dataset. Among the evaluated models, the MLP algorithms achieved the best performance, followed closely by Random Forest and XGBoost. The other algorithms, SVM and Logistic Regression, provided good results, however slightly lower than those of the other methods. For the MLP training, the Adam optimizer was used and built with one single hidden layer with ReLU activation function, an appropriate form for binary classification tasks for cardiovascular risk prediction. Additionally, the paper discusses the complementary role of Internet of Things (IoT) in enhancing data in real time, acquisition, and patient monitoring.

Keywords: Heart disease · machine learning · algorithms · IoT · health prediction

1 Introduction

Nowadays, thanks to recent advancements in healthcare device development and the availability of artificial intelligence (AI), extensive data collection and complex evaluations of medical procedures can be efficiently processed [1]. The systematic analysis of data is essential for improving diagnosis, personalizing treatments, and optimizing clinical outcomes. Moreover, statistical methods help establish a variability in how patients respond to treatments, allowing doctors to find accurate information and make the right decisions even under uncertain conditions.

M. E. Auer et al. (Eds.): IMCL 2025, LNNS 1954, pp. 122–131, 2026.
https://doi.org/10.1007/978-3-032-24998-2_10

According to the current statistics of the World Health Organization (WHO), 20.5 million people lose their lives each year because of cardiovascular diseases [2–5], with rates varying significantly by age and gender [3]. As a matter of prevention, by introducing early-stage education for children, students, and adults on healthcare it can be a strategy for preventing these diseases and, ultimately, an encouragement to adopt long-term healthy and preventive behavior.

From the engineering perspective, it is essential to design and improve cutting-edge technologies to support doctors finding early stages of cardiovascular diseases and, in the meantime, using artificial intelligence (AI) can be classified and identified heart diseases [6] or model cerebral hemodynamics [7]. Moreover, engaging students to contribute scientifically in these research activities represents an opportunity to support their educational development, promote participation in scientific activities, and reinforce their willingness for future professional and academic endeavors.

A few years ago, Modran et al. initiated research studies on this topic by selecting a machine learning (ML) model and identifying performance indicators relevant to the design of Internet of Things (IoT) gadgets [5]. The current study is built and brings additional contributions to that foundational work. Furthermore, the goal is to test additional ML models and to find out which one is the best option to predict if someone might have a risk of heart disease. Additionally, a different data collection with a large volume of data (different from the previous one) is used in this study. The performance parameters identified in this study will support the integration not only in the IoT devices, but also in the mobile medical equipment, thus ensuring scalability and adaptability in educational and healthcare, particularly in rural communities with limited resources and restricted access. This study broadens the scope of [5] by evaluating different MLA-including XGBoost, which was not previously considered. To identify the best predictors, we also conduct thorough feature-importance and correlation analysis. In addition, ML and AI in healthcare, risks and diseases can be detected or prevented in early stages; the diagnosis time could be significantly reduced, so that patients can be continuously monitored with a personalized treatment.

ML is a branch derived from AI that makes it possible for machines to analyze data to perform decision-making and predictions and generate outcomes without specific programmed tasks. As a result, generated classifications of heart diseases [8] (e.g., heart attack, arterial disease and irregular heartbeats, cardiomyopathies, valvular heart diseases, and congenital heart defects) help doctors understand patients' conditions and make faster decisions. Additionally, following the clinical profile of patients, in [8], classifications on acute heart failure are presented. An example of ML and data mining techniques that can be applied in the evaluation and classification can be found in [8].

Medical investigations such as blood work, imaging, genetic screening tests, or vital signs monitoring are only one part between the tests performed in the early discovery of illness and continuous monitoring of patients' health. Additionally, the integration of smart technologies like the IoT has revolutionized patient monitoring. The IoT device allows real-time tracking of vital parameters, enabling early interventions and improving healthcare quality. Associated with Big Data and AI, notably ML, healthcare can be transformed into a proactive, responsive, flexible, and preventive system [9]. Therefore, ML helps not only to manipulate and process large datasets but also to find nuances patterns that would be difficult for humans to see [10–14]. Particularly, ML improves diagnostic accuracy and supports personalized treatment recommendations. Fusing these technologies with real-time monitoring systems can substantially improve efforts to prevent chronic illness and support personalized patient care and treatment [13, 14].

This paper has the following structure: Sect. 2 presents the methodology used in this research. Section 3 describes the implementation steps related to selected algorithms, including Logistic Regression, Support Vector Machine (SVM), K-Nearest Neighbors (KNN), Multilayer Perceptron (MLP), Random Forest and eXtreme Gradient Boosting (XGBoost), followed by data processing, training and evaluation. Section 4 presents experimental results, discussions and comparisons on algorithms, and finally, conclusions are summarized.

2 Methodology

The methodology applied in this study explore the possibilities to establish a comparative approach between several ML algorithms generally used for predicting the exposure to various cardiac disorders, such as coronary artery disease, heart failure and cardiac arrhythmias, cardiomyopathies, valvular heart diseases, and congenital heart defects. Therefore, the focus is directed towards identifying the most efficient classification techniques, taking into consideration the medical context and the high level of accuracy required for diagnosis. Thus, the methodological strategy is defined by following four steps:

(i) Step 1: Define the objective.
(ii) Step 2: Choose the most appropriate algorithm,
(iii) Step 3: Develop and implement a representative dataset.
(iv) Step 4: Apply evaluation metrics for the performance of the model.

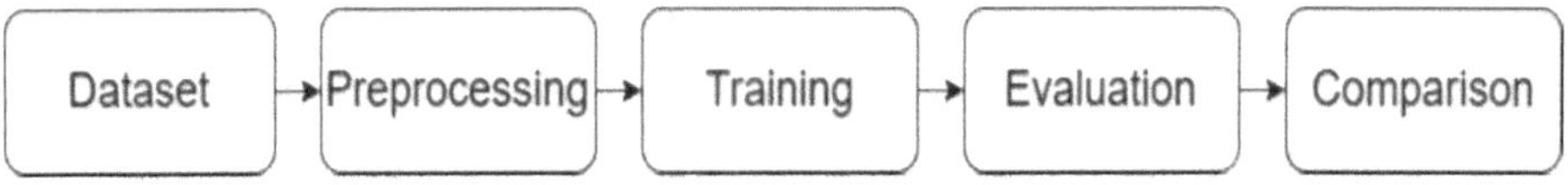

Fig. 1. Implementation steps applied to a dataset of 1026 patients with 14 health features.

Figure 1 describes the core stages of the approach used in this research. The process begins with dataset selection (e.g., heart disease dataset), followed by a pre-processing phase where missing values and normalization for specific algorithms are handled. Subsequently, multiple ML models are trained. Each model is then evaluated using standard performance metrics [8, 15] (e.g., precision, recall, F1-Score, accuracy). Finally, the outcomes are examined to ascertain the most effective model.

In data science, several ML algorithms are frequently chosen for this type of problem. Some of the most common can be enumerated as follows: Logistic Regression, Support Vector Machine (SVM), K-Nearest Neighbours (KNN), Random Forest, XGBoost, K-Means Clustering, and Multilayer Perceptron (MLP) with a brief description provided in [16]. The first five algorithms are supervised learning methods; meanwhile,

K-Means is an unsupervised algorithm. This section allows us to highlight the differences between supervised and unsupervised classification techniques, emphasizing the models' capability to distinguish between patient groups based on clinical features.

The parameters for evaluation that were employed include precision, recall, F1-score, and overall accuracy [16, 17]. These values are extracted and evaluated for each ML model to identify which algorithm is the most stable and efficient, considering the sensitive medical context. This approach seeks to maintain a balance between the interpretability of the models and their predictive performance, providing a transparent perspective on the usefulness of each model for the analyzed problem.

3 Study Implementation

This section presents technical details in order to implement the proposed solution, since we process data. Configure models, as well as train them based on the evaluation methodology. The dataset processing steps are described along with the cross-validation strategy and performance evaluation metrics. These aspects are what the results need. They also require making a comparison of the results.

The dataset that was used within this study was obtained from an open-source repository (from Kaggle, [18]), and it contained pre-collected clinical data for cardiovascular patients. Although we did not do the data collection, we carefully reviewed its structure plus quality to ensuring it could train machine learning models. The dataset includes anonymized clinical records. These data were originally collected from within medical institutions. They were made available publicly for educational and research purposes.

3.1 Implementation Specifics

Clear steps were followed to evaluate selected ML algorithms able to cover a range of supervised and pattern discovery techniques. Aspects and comparisons of these algorithms are given in [19–21, 23]. These algorithms were tested on the heart disease dataset, so that results could be compared based on how accurately and efficiently they predict the risk of disease [22].

For the implementation, the Python programming language and the Scikit-learn library were used for Machine Learning tasks. Additionally, the Pandas library was included for data processing, and NumPy was applied for computational tasks. Matplotlib and Seaborn were used for data visualization, as well as for a better understanding of the distributions of variables and the relationships between them.

Analysing the ML algorithms' characteristics, each of them provides essential information. For example, the Logistic Regression algorithm is a direct method particularly used for binary classification, and is well fitted to identify if a patient presents any cardiovascular disease. Further, the SVM is able to detect an optimal hyperplane that can divide classes, even when the data is non-linear. When examining the KNN algorithm, this one is able to classify a sample by identifying the majority class among the closest points and is advantageous when data from patients is described by a complex, non-linear relationship. An artificial neural network that uses feedforward technology and is known as the MLP models, that corresponds to non-linear relationships based on one or more hidden layers. When feature interactions are difficult for conventional models to capture, it is very helpful. The MLP model in this paper used the Adam optimizer for training and was constructed with a single hidden layer with ReLU activation, which makes it appropriate for binary classification tasks like cardiovascular risk prediction. Two additional ML algorithms that strike interest are the Random Forest and XGBoost, methods that supports the classification accuracy by combining multiple decision trees. Therefore, the Random Forest and XGBoost models are robust and handle complex datasets well.

3.2 Data Processing

Data processing was an essential step before training the models. The dataset consisted of over 1000 observations and 14 variables, which included clinical features like age, sex, blood pressure, cholesterol levels, electrocardiographic results, and more. Each feature that was there was examined with care. Its quality and consistency were ensured by this careful examination.

This study used a dataset of 1000 patient records, each containing 14 clinical characteristics including height, weight, systolic and diastolic blood pressure, cholesterol, glucose levels, physical activity, smoking and alcohol use, age, sex, and height. Six ML methods were employed to identify and classify eventual heart risks:

(1) the Logistic Regression
(2) the Support Vector Machine (SVM)
(3) the K-Nearest Neighbors (KNN)
(4) Multilayer Perceptron (MLP)
(5) Random Forest
(6) XGBoost

These models were trained and tested using Python and Scikit-learn libraries. To evaluate, verify, and classify the influence of each model, the accuracy, precision, recall, and F1-score metrics were included.

Before applying the ML algorithms, the data was filtered by checking for empty entries. If a value was missing, it was imputed using mean imputation for continuous variables. In cases where imputation wasn't suitable, rows with missing values were removed to maintain the dataset's integrity.

For algorithms that happen to be sensitive to that scale of data, such as Logistic Regression, SVM, and KNN, we normalize by way of the StandardScaler method. This process changes features so that they possess a mean of zero and also a standard deviation equal to one, which standardizes the data. This also guarantees each feature contributes the same to model performance. Because these tree-based models are not sensitive to feature scaling, Random Forest, along with XGBoost, were not normalized. A split of 80% to 20% was used for segmenting the dataset into training and testing sets. This division ensures that a separate unseen set exists for model testing and training on data. Models were trained on the training set. The testing set assessed their performance during.

3.3 Training and Evaluation

Once the data was preprocessed, training and assessment of the models followed. The metrics for evaluation used to examine the model's performance were precision, recall, F1-score, and overall accuracy. These metrics apply specifically to medical applications. False positives as well as false negatives must be balanced in these critical applications.

The training process was conducted using Scikit-learn in Python, where the models were fitted to the training data, and the testing set was used to evaluate their performance. Cross-validation was effectuated to insure that the models were not overfitting to the training data and could perform robustly across contexts well to new, unseen data. Repeated training was likewise used to train the MLP model, and the same metrics were used to assess it. It achieved competitive performance and successfully captured intricate, non-linear patterns.

4 Experimental Results

To evaluate the models, we trained and tested Logistic Regression, SVM, KNN, Random Forest, and XGBoost on the heart disease dataset. Each model underwent performance evaluation using precision, recall, F1-score, and overall accuracy.

Table 1 presents the classification report for each model, including scores for both classes (presence or absence of cardiovascular risk). It can be observed that the MLP Classifier model succeeded in the highest F1-score for both classes, followed by Random Forest and XGBoost. SVM also achieved good scores, whereas Logistic Regression showed slightly lower performance relative to the other models.

The models were trained by using the Scikit-learn library, with StandardScaler applied for Logistic Regression, SVM, and KNN; Random Forest and XGBoost were trained without normalization. Training was using 80% for that train-test split, as testing used 20%.

On the whole accuracy of each model is summarized. MLP obtained an accuracy of 99%, the highest. Random Forest and XGBoost followed next. SVM achieved 88% overall accuracy, and KNN got around 83%. However, Logistic Regression only achieves about 79%. MLP, Random Forest, and XGBoost present efficiency in this kind of medical classification problem, displaying ensemble methods' robustness, like SVM and KNN. We can choose the best predictive strategy for the early detection of cardiovascular diseases through the testing of different approaches.

Figure 2 shows the confusion matrix for the MLP model, where there is revealed that the model classified 102 patients without cardiovascular disease (known as true negatives) and 100 patients with the disease (known as true positives). Only 3 patients with the disease were missed by the model (false negatives). With slightly more false negatives than false positives, this distribution points to a balanced performance, which is crucial in medical settings where failing to diagnose a patient can have dire repercussions. All things considered, the MLP model exhibits good generalization and precise classification for both classes.

Table 1. Classification Report based on standard performance metrics.

Model	Class	Precision	Recall	F1-Score	Accuracy
Logistic Regression	0	0.85	0.72	0.78	0.79
	1	0.76	0.87	0.81	
SVM	0	0.93	0.83	0.88	0.88
	1	0.85	0.94	0.89	
K-Nearest Neighbors	0	0.88	0.77	0.82	0.83
	1	0.80	0.89	0.84	
MLP Classifier	0	0.97	1.00	0.99	0.99
	1	1.00	0.97	0.99	
Random Forest	0	0.97	1.00	0.99	0.98
	1	1.00	0.97	0.99	
XGBoost	0	0.97	1.00	0.99	0.98
	1	1.00	0.97	0.99	

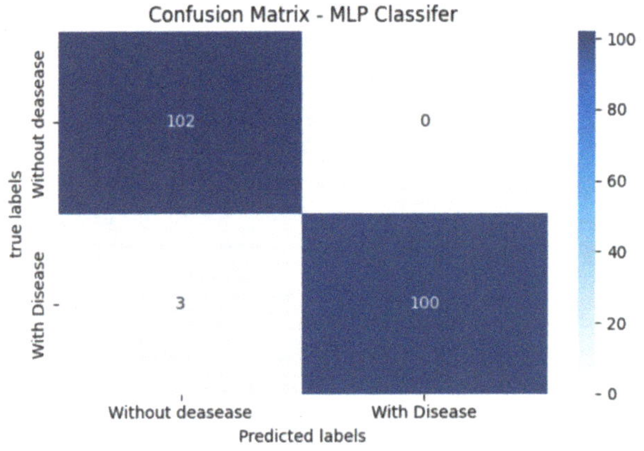

Fig. 2. MLP Classifier confusion matrix.

5 Discussions and Conclusions

This research showed that machine learning algorithms can accurately forecast cardiovascular disease risks through clinical data analysis. MLP produced the highest performance results while Random Forest and XGBoost followed closely behind. The results from SVM and Logistic Regression were good, but not as strong as the other methods.

IoT technologies improve patient health monitoring effectiveness by collecting and analyzing data in real time. The integration of predictive models with smart health infrastructures results in improved healthcare services and better quality of life for patients.

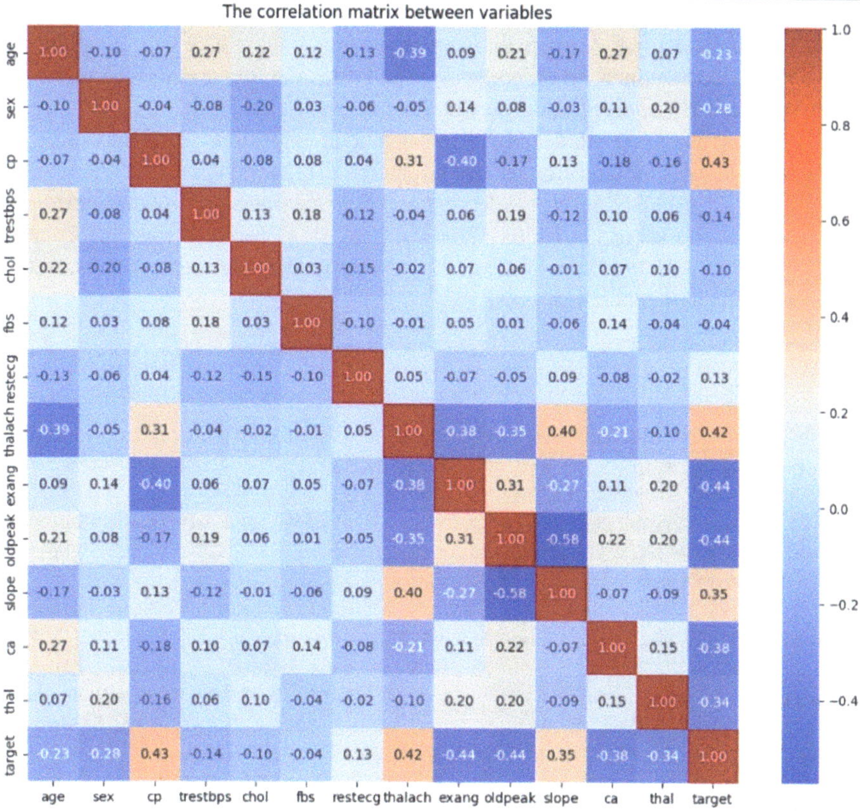

Fig. 3. Correlation Matrix.

A Pearson correlation matrix was computed to assess the linear relationships among the 14 clinical features, excluding the target variable, in the updated heart disease dataset. The primary goal was to detect which variables show the strongest association with the target variable (presence or absence of cardiovascular disease). As shown in Fig. 3, chest pain type (cp) demonstrates the strongest positive correlation with the disease outcome ($r = 0.43$), followed by the greatest heart rate achieved ($r = 0.42$) and tilt of the ST segment ($r = 0.35$). These variables are predicted to contribute meaningfully to anticipate cardiovascular conditions.

In contrast, we found some significant negative correlations between the CAD and the exercise-induced angina ($r = -0.44$), ST depression induced by exercise ($r = -0.44$), major CESG vessels colored ($r = -0.38$), and thalassemia ($r = -0.34$). These negative connections reveal that patients with higher levels of these features are less likely to have heart disease, or that these features are protective or inverse risk factors.

This feature correlation analysis provides a picture of which clinical predictors-such as chest pain type and maximum heart rate, are most tied to risk for cardiovascular disease. These insights can be used to inform the selection of features in subsequent predictive models by prioritizing the most important measurements, and thus improve interpretability and computational efficiency. The emphasis on highly correlated features allows models to preserve predictive performance with reduced input and this also reduces the requirements for data collection (and, thus deployment in IoT-based monitoring systems). By concentrating on the most relevant clinical features (e.g., type of chest pain, peak heart rate), this study provides guidance for future models. Future deep learning initiatives or methods for selecting features in bigger datasets may benefit from these findings. Automated feature selection and the integration of bigger, more varied datasets will be the subjects of future studies. Next steps could involve incorporating deep learning models and the adoption of bigger and more heterogeneous datasets for better prediction and generalization.

References

1. Bajwa, J., et al.: Artificial intelligence in healthcare: transforming the practice of medicine. Future Healthc. J. **8**(2):e188-e194 (2021). https://doi.org/10.7861/fhj.2021-0095. PMID: 34286183; PMCID: PMC8285156
2. World Heart Observatory | Global Cardiovascular Health Insights. Last accessed 18 June 2025
3. https://ourworldindata.org/grapher/cardiovascular-disease-death-rate-age-group-who-mdb. Last accessed 18 June 2025
4. Lopez, O., et al.: Cardiovascular disease. In: StatPearls [Internet]. Treasure Island (FL): StatPearls Publishing (2025). Available from: https://www.ncbi.nlm.nih.gov/books/NBK535419/
5. Modran, H.A., et al.: Artificial intelligence system for predicting cardiovascular diseases using IoT devices and virtual instrumentation. Lecture Notes in Networks and Systems, vol. 298. Springer, Cham (2022). https://doi.org/10.1007/978-3-030-82529-4_28
6. Abbas, S., et al.: AI framework for heart disease classification from audio signals. Sci. Rep. **14**, 3123 (2024)
7. Bogdan, I.C., et al.: A study of mathematical models used in cerebral hemodynamics circulation, Polytechnic University of Bucharest, Scientific Bulletin, Series C, Vol. 85, Iss. 3 (2023). ISSN 2286-3540
8. Garbacea, A.S., et al.: Machine learning and data mining techniques for detecting fraudulent job postings. The 22nd Int. Conf. on Smart Tech. & Education (2025)
9. González-Pacheco, H., et al.: Clinical phenotypes, aetiologies, management, and mortality in acute heart failure: a single-institution study in Latin-America, ESC Heart Failure 2021; 8: 423–437 (2021). Pub. online 2020 in Wiley Online, https://doi.org/10.1002/ehf2.13092
10. Udegbe, F.C., et al.: The role of artificial intelligence in healthcare: a systematic review of applications and challenges. Int. Med. Sci. Res. J. **4**(4) (2024). https://doi.org/10.51594/imsrj.v4i4.1052
11. Ciubotaru, B.-I., et al.: Prototype results of an IoT system using wearables and artificial intelligence for the detection of frailty in elderly people. App. Sciences, MDPI (2023)
12. Shehab, H.H., et al.: Developing an AI model that relies on mobile health devices to track heart activity. J. Tech. (2024). https://doi.org/10.51173/jt.v6i3.2576
13. Sanap, A. R. et al.: Predicting Heart diseases in IoT-Based electronic health records: a federated learning approach. **20**(2s), 1132–1144 (2024). https://doi.org/10.52783/jes.1759
14. Shukla, S., et al.: Predicting cardiovascular disease through IoT and deep learning methods. 250–255 (2024). https://doi.org/10.1109/icses63445.2024.10763287
15. Sathe, B.Y., et al.: IoT-Enabled heart disease prediction: harnessing machine learning for proactive healthcare. 1–8 (2024). https://doi.org/10.1109/iccubea61740.2024.10775150

16. Schlosser, T., et al.: A consolidated overview of evaluation and performance metrics for machine learning and computer vision (2024). https://doi.org/10.13140/RG.2.2.14331.69928
17. Sengupta, J.: Investigation of deep learning models for analysis of heart disorders in smart health care based IoT environment. J. Smart Internet Things **1**, 1–16 (2024). https://doi.org/10.2478/jsiot-2024-0001(2024)
18. https://www.kaggle.com/datasets/juledz/heart-attack-prediction?resource=download. Last accessed 18 June 2025
19. Machine Learning Algorithms You Should Know for Data Science-StrataScratch. Last accessed 2025 June 18
20. Akther, K., et al.: Multi-Faceted approach to cardiovascular risk assessment by utilizing predictive machine learning and clinical data in a unified web platform. IEEE Access **12**, 120454–120473 (2024). https://doi.org/10.1109/ACCESS.2024.3436020
21. Pardhu, T.: Leveraging IoT and machine learning for improved health prediction systems. In: P.T. (Ed.), Health Prediction Systems (pp. 278–306). IGI Global (2024). https://doi.org/10.4018/979-8-3693-1186-8.ch016
22. Jenifer, A., et al.: Edge-based heart disease prediction using federated learning. 294–299 (2024). https://doi.org/10.1109/icc-robins60238.2024.10534005
23. Pandey, V.K., et al.: An efficient and secure framework for smart healthcare using IoT and machine learning. 1–5 (2024). https://doi.org/10.1109/dasa63652.2024.10836569

Gamified Radar Simulation: Augmented Reality Approaches for Military Education

Melina Iftime and Ecaterina Liliana Miron(✉)

Henri Coanda Air Force Academy, Brasov, Romania
{melina.iftime.std,liliana.miron}@afahc.ro

Abstract. This paper presents a summary of an independent educational initiative, Radar System Simulation in Augmented Reality. The project was launched as an internal experiment, developed by students specializing in airspace surveillance. In its initial version, the experiment aimed to project a radar system in augmented reality, to facilitate understanding of its components and operational principles. However, the lack of a compelling visual impact led to a redefinition of the objective, shifting the focus toward gamifying the radar operation process for educational purposes. The application was developed in Unreal Engine, using the Blueprint visual scripting system to simulate the radar beam, target detection, and automatic responses. For testing, the application was connected to HoloLens 2 via the Holographic Remoting Player app. This paper outlines the development methodology, along with the results obtained, the technical limitations encountered (including challenges with native deployment on HoloLens), and the potential of this approach for practical training in the technical and military fields.

Keywords: augmented reality · gamification · radar · Unreal Engine

1 Introduction

Surveillance systems are advanced technological platforms designed to detect and track aerial or maritime objects such as aircraft, ballistic missiles, or ships. Diverse radar installations are used across military and civilian infrastructures, from airspace surveillance to coastal and airborne radars, covering altitudes between 500 and 30,000 m for multidirectional monitoring.

Training operators for these systems typically requires costly equipment and operational conditions that are difficult to replicate in academic or military institutions. These constraints reduce the frequency and depth of hands-on exercises. Augmented reality (AR) offers a scalable and interactive alternative that enables simulation-based training without a physical infrastructure [1].

To address these challenges, this project proposes an interactive educational application built into Unreal Engine, aimed at simulating the detection logic of a surveillance radar. The application illustrates how a radar scans, identifies, and tracks virtual objects

M. E. Auer et al. (Eds.): IMCL 2025, LNNS 1954, pp. 132–143, 2026.
https://doi.org/10.1007/978-3-032-24998-2_11

in a defined airspace, based on a logic that emulates radar beam movement, target trajectories, and configurable system responses [2].

The Unreal Engine platform allows the creation of dynamic, game-like scenarios that support immersive, contextualized learning. Each student can customize their own training missions by configuring radar parameters and target positions, encouraging engagement, personalization, and adaptive learning.

This approach aligns with the goals of the AILA-25 section by integrating advanced simulation technology to support Authentic Learning and Personalized Learning. Given its development by students in the military education domain, this project demonstrates how AR and game engines can facilitate realistic and accessible training in sensitive technical fields.

1.1 Why Gamification and Augmented Reality

Integrating gamification and augmented reality into technical and military education offers an innovative response to the challenges of limited access to real radar systems and operational equipment. The proposed application introduced several unique elements.

a) Visualization of radar behavior in 2D or 3D enables students to observe the scanning and detection processes in real time.
b) Customizable training scenarios in which students can modify target positions and radar parameters and simulate tactical responses.
c) Interactive learning model, based on real-time feedback and decision-making in an augmented reality environment.
d) The use of Unreal Engine Blueprints allows students to understand both system logic and visual behavior without requiring advanced coding skills.
e) Partial deployment on HoloLens 2, offering immersive visualization and alignment with current trends in military and aerospace simulations.

These features contribute to a shift from passive content consumption to active learning experiences, supporting both Authentic Learning and Personalized Learning.

2 Literature Review. Radar Systems Theory and Interactive Technologies

The integration of educational mini-games into military academic curricula serves two main purposes: (1) enhancing the understanding of complex technical system, such as radar, by enabling students to simulate and manipulate their structure and core parameters, and (2) fostering digital skills and intellectual autonomy through creative and interactive problem-solving activities.

To validate this approach, relevant research on radar simulation and the use of interactive platforms, especially game engines like Unreal Engine, must be considered. Commercial tools such as ScanView [3], RadarSimX [4], and SatGen [5] offer high-fidelity radar/GNSS simulations but are often cost-prohibitive, non-interactive, and not designed for educational use.

In contrast, flexible platforms such as Unreal Engine support personalized, student-centered learning scenarios. The proposed application demonstrates how students can apply technical knowledge through simulation, scenario design, and direct engagement alignment with the AILA-25 focus on interactive and adaptive learning.

The Unreal Engine, which was initially developed for video games, is now widely used in education and simulation. Projects such as Meta Immersive Synthetics, Flight-Forge, and AirSim have demonstrated their value in aviation, robotics, and AI training environments. Platforms such as HEROES allow immersive emergency response simulations, whereas AirSim [6] supports AI and autonomous vehicle research.

These examples show that Unreal Engine is well-suited for developing authentic learning environments, combining interactivity, technical fidelity, and scalability.

The radar simulation created in this project was projected onto HoloLens 2, which allowed students to visualize and interact with radar elements in their real surroundings. The mobility and infrastructure-free nature of the headset support personalized learning, bridging the gap between theory and applied military training.

3 Proposed Methodology

The application was designed to simulate the basic principles of a surveillance radar system in an interactive format. The main goal was to visually and behaviorally represent how a radar detects and tracks objects entering a defined airspace. The educational scenario allows the user to configure the position of the radar and targets for different system operations and response hypotheses.

As previously mentioned, the Unreal Engine was chosen as the development platform because of its capabilities for image creation, visual logic implementation through Blueprints, and built-in support for augmented reality. The application architecture includes the following components:

Radar Model: Composed of a static object placed in the scene and an animated rotating beam that simulates the radar's detection angle.

Detectable Objects: Objects simulating aerial targets (aircraft), manually or automatically placed within a 2D or 3D virtual space.

Detection Logic: Implemented using collision and visual triggers; when the radar beam intersects a target, the target changes color and a response is triggered at the interface.

User Interface: Adapted for eye tracking and gesture control, allowing real-time manipulation of the positions of both radar and targets.

The application was developed using the visual programming system Blueprints, with the radar logic controlled through conditional functions and customized animation cycles.

For deploying the application on the HoloLens 2 device, official packages provided by Microsoft and Epic Games were used (Microsoft Mixed Reality Toolkit and the HoloLens plugin for Unreal Engine).

The integration process with augmented reality involves:

- The size and positioning of the virtual objects are adjusted to match the constraints of the physical environment in which the HoloLens is used.
- Implementing gestures and gaze tracking controls allow students to interact with the simulation naturally and without external controllers.
- Testing the application on the HoloLens headset, followed by deployment on the actual device to calibrate object positioning and optimize graphical performance.

By constructing a radar scenario in an interactive 3D environment that overlaps the physical space using HoloLens 2, the student benefits from a clear and immersive learning experience. They can observe how the radar beam operates, adjust the function of targets, and directly observe the system's reactions, facilitating a better understanding of how theoretical principles are applied in realistic situations.

The Logic Flow for Radar Simulation using Unreal Engine and HoloLens 2 is illustrated in Fig. 1.

The implementation of the simulation logic flow, as illustrated in Fig. 1, enabled the concrete development of the application, which was built in the Unreal Engine and adapted to operate in an augmented reality environment. The implementation is structured in two main stages: development of the radar's functional logic and integration of the application with the HoloLens 2 device.

3.1 Functional Logic and Radar Simulation

The radar beam is visually represented as a rotating transparent plane that scans a defined field of view, mimicking the operation of a real surveillance radar. In the Unreal Engine, this beam is implemented as a visual effect that rotates around the radar vertical axis.

The described logic is summarized in the following pseudocode, which outlines the sequence of steps and system behavior during radar simulation (see Fig. 2).

The pseudocode implements the logic flow illustrated in Fig. 1. This correspondence is essential to understand how a complex radar scenario can be translated into an interactive educational application.

Providing and using the pseudocode helps to plan and explain the logic of the application to students. This step serves as the foundation for the later integration of the project into an augmented reality device such as HoloLens 2.

3.2 Application Integration with HoloLens 2

Figure 3 provides a graphical representation of the decision and technical flow for integrating the Unreal Engine application into HoloLens 2, highlighting the connection between the educational objectives and the final implemented solution.

The initial goal of integrating the radar application into HoloLens 2 was to transpose the interactive simulation into an augmented reality environment, where the student could visualize and control the radar detection process in real-time using head movements, voice commands, or gestures. This approach aimed to provide an educational experience in which radar elements (the beam, targets, and detection markers) are anchored in a real physical environment, thereby simulating real tactical conditions in a controlled manner.

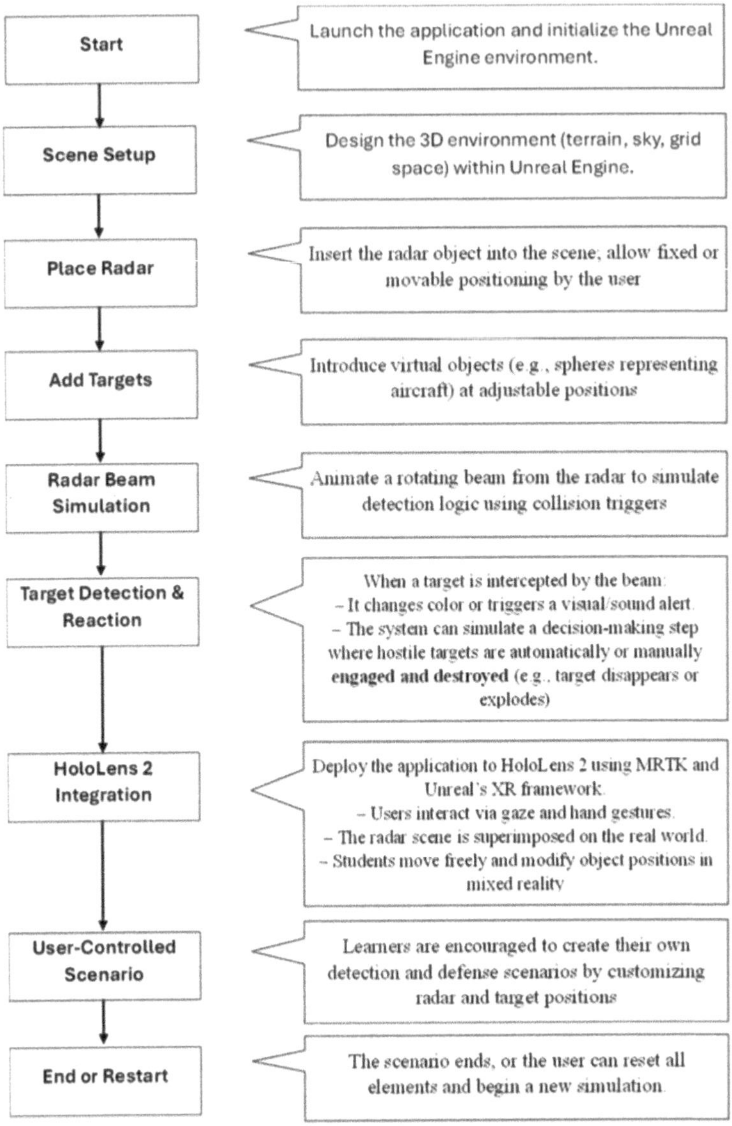

Fig. 1. Logic flow for radar simulation using Unreal Engine and HoloLens 2

4 Implementation of the Radar Simulation Application

The technical limitations encountered in achieving this objective were mainly related to the complexity of configuring the Unreal Engine environment for HoloLens devices. Although Unreal Engine provides support for augmented reality, full integration requires additional modules, compatibility with the Mixed Reality Toolkit (MRTK), adaptation of the interface to gesture and gaze control, and optimization of the limited resources of HoloLens 2. These requirements demanded an advanced level of XR development, which was beyond the immediate scope of this project. Fully transferring the simulated logic—including complex Blueprints and radar animations—into AR format proved technically challenging during the project.

```
PROCEDURE StartSimulation
  // Step 1: Initialize environment
  InitializeScene()
  LoadRadarObject()
  // Step 2: Set radar parameters
  SET detectionRange ← value
  SET rotationAngle ← value
  SET rotationSpeed ← value
  // Step 3: Create visual beam
  radarBeam ← CreateMeshOrDecal()
  AttachToRadar(radarBeam)
  // Step 4: Generate virtual targets
  FOR EACH target IN TargetList DO
    SET target.initialPosition ← position
    SET target.speed ← predefinedValue
    ALLOW userToModify(target.speed, target.position)
    SynchronizeWithRadarDisplay(target)
  // Step 5: Begin rotation and detection loop
  WHILE simulationIsRunning DO
    RotateBeam(radarBeam, rotationSpeed)
    UpdateTargetPositions()
    FOR EACH target IN TargetList DO
      IF BeamOverlapsTarget(radarBeam, target) THEN
        MarkAsDetected(target)
        LogDetection(target)
        IF target.type = "hostile" THEN
          TriggerNeutralization(target)
        END IF
      END IF
    END FOR
  END WHILE
  // Step 6: Reset state if needed
  IF ResetTriggered THEN
    ResetAllObjectsToInitialState()
  END IF
END PROCEDURE
```

Fig. 2. Pseudocode – Radar beam simulation

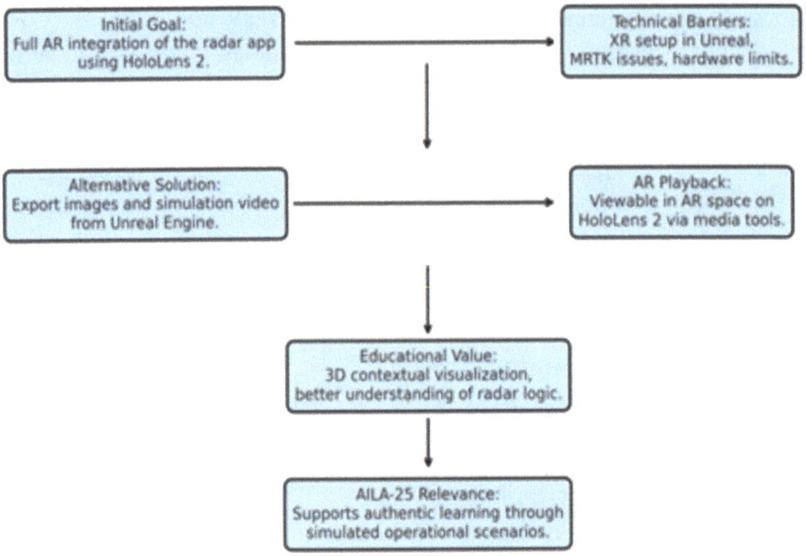

Fig. 3. HoloLens integration diagram

The practical solution implemented to maintain the augmented reality component was to export relevant images from the radar simulation and create a short demonstration video showcasing the application's functionality in Unreal Engine [6]. These materials were later played directly on the HoloLens 2 using the device's standard media playback functions. Thus, even though the application does not run natively, the experience is partially transposed into augmented reality, enabling a form of visual immersion.

The educational value of this alternative solution lies in the fact that it provides students with the opportunity to visualize radar operation in a real-world space and associate theoretical learning with a concrete visual experience. The presentation of the radar beam, detections, and target movements in an AR environment supports the construction of an intuitive mental model of radar functionality.

The correlation with the overall objective of the project remains valid: by enabling three-dimensional visualization of the radar process in an augmented context, the goal of transforming theoretical content into a practical, reality-based experience is achieved. Even in its current form, this application contributes to interactive learning and the development of tactical thinking specific to military scenarios.

5 Results of the Application Implementation

The results of implementing the radar application developed in Unreal Engine and visualized through HoloLens 2 reflect the achievement of the key design and simulation objectives. The student can observe the behavior of the radar, the target detection, and automatically trigger a tactical response. The project was tested using HoloLens via the Holographic Remoting Player application, which connects wirelessly to the laptop by entering the network IP, thus enabling the real-time rendering of the Unreal scene in augmented reality.

The development of the process led to a functional simulation that combined the technical principles of radar systems with interactive elements and visual support in an AR environment. The following section presents the results in the form of visual captures from Unreal Engine and images from the projection rendered through the HoloLens 2.

5.1 Radar Scanning Mechanism

The radar was implemented using an animated visual beam (scanner), created using dynamic materials. Texture and animation give the impression of rotation, represented by a transparent plane that scans the 3D space. Detection is achieved through collision between the beam and target (Fig. 4).

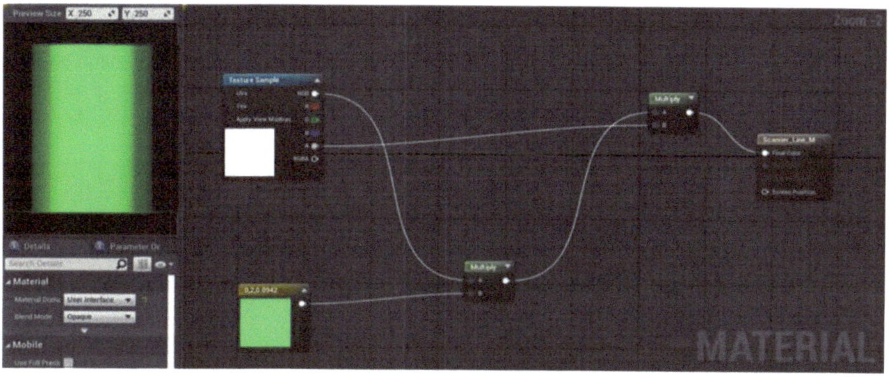

Fig. 4. Material and structure for the radar beam. [from Melina Iftime thesis, Augmented reality radar system simulation, Henri Coanda Air Force Academy, 2025]

5.2 Functional Parameters and Radar Logic

The operating parameters, such as the detection range, coverage angle, and rotation speed, were controlled through the Blueprint system in the Unreal Engine. These parameters can be adjusted by the student to observe their impact on radar performance (Fig. 5).

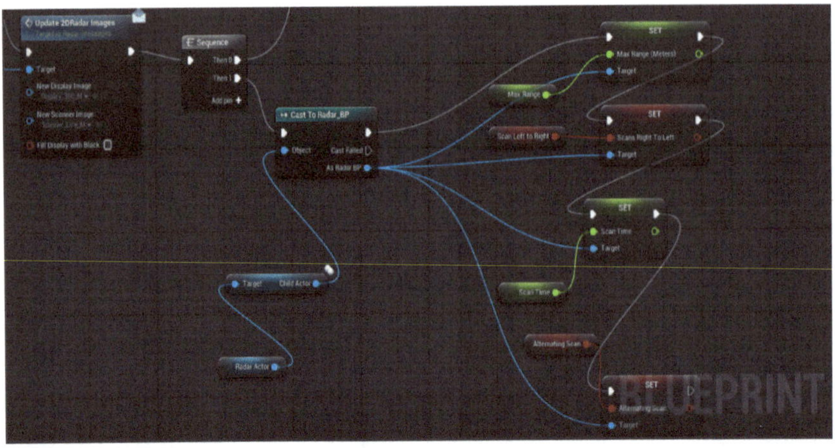

Fig. 5. Radar parameter configuration in Blueprint. [from Melina Iftime thesis, Augmented reality radar system simulation, Henri Coanda Air Force Academy, 2025]

5.3 Representation of the Radar and Targets

The radar is visually represented through a rotating "radar scope" display, in which a green scanning beam continuously sweeps the defined space. The targets are dynamic 3D objects that move and appear on the screen when detected by the radar (Fig. 6).

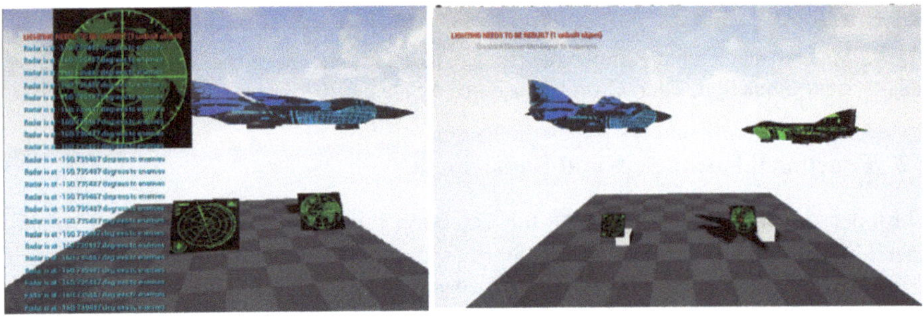

Fig. 6. Radar and target implementation. Radar display with the positioning of the detected target [from Melina Iftime thesis, Augmented reality radar system simulation, Henri Coanda Air Force Academy, 2025]

Upon target detection, the application generates visual feedback; the radar beam may change color, a tracking line toward the target is displayed, and in some cases, an animation simulating the neutralization of the target is triggered. These responses enhance the understanding of radar operations and increase the applied and interactive nature of the simulation.

5.4 Application Integration with HoloLens 2

The Holographic Remoting Player app was used to visualize the application in AR. This application enables wireless transmission of the Unreal Engine scene to HoloLens 2. The computer's IP address was entered into the HoloLens device, and the simulation streamed live from Unreal.

Below, Figs. 7 and 8, are screenshots from the application viewed through HoloLens 2 using Holographic Remoting Player. These images illustrate how radar elements and targets are projected onto the user's surrounding physical space.

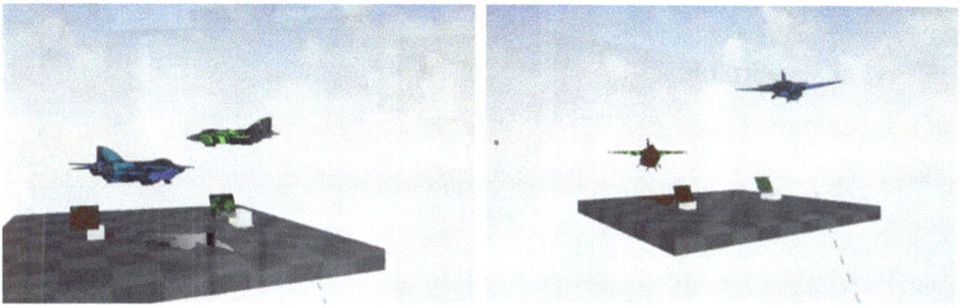

Fig. 7. Projection on HoloLens 2 [from Melina Iftime thesis, Augmented reality radar system simulation, Henri Coanda Air Force Academy, 2025]

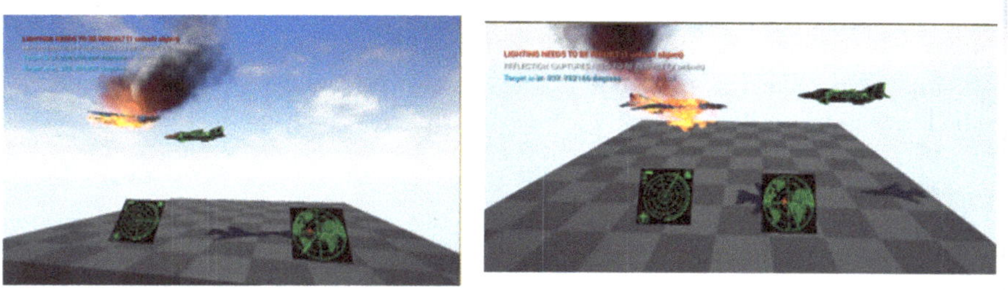

Fig. 8. Visualization of the decision to neutralize the enemy object projected on HoloLens 2.

6 Preliminary Evaluation and Pilot Scenario

Although the application is still in the development and experimental testing phases, pilot scenarios have been conducted to verify the basic functionality of the radar logic and the visual interaction mechanisms in simulated and augmented environments.

Preliminary tests were carried out in a controlled environment, using a workstation with an Unreal Engine and HoloLens 2 headset. The users (three final-year students specializing in aerial surveillance) interacted with the application by following a predefined set of instructions. The activities included the following.

- placing targets within the virtual space;
- observing the radar's reaction upon detecting them;

- tracking the visual changes on the interface;
- visualizing the simulation on HoloLens 2 through video playback.

At an early stage, the evaluation has not yet relied on standardized quantitative metrics but has focused on the following.

- radar display accuracy (consistency of target positioning on the screen in relation to their physical/virtual location);
- student response (directly observed: interest, clarity, ability to anticipate system behavior);
- level of interaction and visual feedback at the time of detection (color changes, visual effects, etc.).

6.1 Preliminary Results

Following the initial evaluation, several observations were made:

1) The radar system successfully detected objects entering the scanning beam, thereby triggering appropriate visual feedback.
2) Students appreciated the way the application helped them understand radar functionality through an intuitive and interactive interface.
3) Although native playback on HoloLens was technically limited, the experience was preserved through video exports, enabling contextualized augmented reality visualization.
4) An automatic tactical response to target detection was a key feature observed during testing. When an enemy object is identified, the system initiates a visual response that simulates neutralization (e.g., color change, disappearance, and destruction animation). This added layer of interactivity reinforced tactical thinking and enriched the learning process.

6.2 Ongoing Work

In the next phase, extended testing is planned with larger groups of students, using evaluation questionnaires, performance metrics, and measurements of learning progress. The following aspects were monitored.

- reaction time in dynamic scenarios;
- ability to configure one's own radar scenario;
- level of understanding gained after using the application.

7 Discussions, Conclusions, and Future Directions

7.1 Discussions

The radar application developed in Unreal Engine and partially tested through HoloLens 2 successfully conveyed the essential concepts of radar system functionality in a visual and interactive manner. The rotation of the beam, target detection, and automatic response in a "target elimination" scenario were functionally implemented. Detection was rendered correctly using collision triggers, and the display of visual feedback at the moment of interception supported intuitive learning.

The application supports Authentic Learning by allowing students to observe complex technical processes in an adaptable environment. Through the integration of gamification elements (custom scenarios, automatic reactions, dynamic animations), students can experiment, explore, and learn actively. Each student can configure their own scenario by choosing the number and position of targets or adjusting radar parameters, thus contributing to personalized learning.

The current limitations of the project are both technical and methodological.

Owing to incomplete compatibility between the Unreal Engine and the HoloLens 2 environment, full native playback could not be achieved. The practical solution implemented was exporting the simulation as a video or in the form of screen captures. Other challenges include the wireless streaming latency, need for spatial calibration to ensure accurate tracking, and network security restrictions in institutional environments. Additionally, time was a significant constraint, as the experiment could not be fully completed before the thesis submission deadline.

7.2 Conclusions and Future Directions

The main achievement lies in the creation of an interactive educational application that simulates the functioning of an air surveillance radar system, integrating visual logic, decision-making components, and potential for expansion into augmented reality. The developed interface allows the testing of different radar behaviors and real-time observation of the system's reaction to virtual threats.

The relevance of the application in technical and military training is significant for fields such as air defense, tactical surveillance, or the preparation of personnel for real radar systems. By using simulated scenarios, students can learn the basic principles without operational risks or high costs.

Future developments aim to integrate AI algorithms capable of generating automatic behaviors such as evasion, target coordination, or defensive response. Additionally, features for analyzing student performance will be added, along with full integration with HoloLens or even META Quest 3, and the expansion of testing to larger student groups to more accurately evaluate student satisfaction.

References

1. Ocaña, M., et al.: Are VR and AR Really Viable in Military Education? A Position Paper. In: Developments and Advances in Defense and Security, Smart Innovation, Systems and Technologies, 165–177. Springer, Singapore (2023)
2. Skolnik, M.I.: Introduction to Radar Systems, 3rd edn. McGraw-Hill, New York, NY, USA (2005)
3. https://dspnor.com/products/scanview-framework/?utm_source=chatgpt.com. Accessed between January 10 and June 15 2025
4. https://radarsimx.com. Accessed between January 10 and June 15 2025
5. https://www.labsat.co.uk/index.php/en/products/satgen-simulator-software?utm_source=chatgpt.com. Accessed between January 10 and June 15 2025
6. https://www.unrealengine.com/en-US. Accessed between January 10 and June 15 2025

Tracing the Formation of Modality in Music Within the Indo-European Broadcasting Continuum Using AI: The Eastern Mediterranean in Perspective

Anastasios Nikiforos[1], Dionysios Politis[1(✉)], and Nektarios Paris[2]

[1] School of Informatics, Aristotle University of Thessaloniki, Thessaloniki, Greece
dpolitis@csd.auth.gr

[2] Dept. of Music Science and Art, University of Macedonia, Thessaloniki, Greece

Abstract. Modal Music has gained global incentive by broadcasts either in the conventional form (radio-television and satellite emissions) or by multimedia modules incorporated within various Information and Communications Technologies (ICTs) modules. The second conduit of dissemination offers, furthermore, extra tools and channels for Cultural Reconstructions, associating Learning Architectures with AI agent modelling. As a result, the Indo-European musical continuum has recently revived through Content Management Systems, Digital Libraries and Animated Agents offering Deep Learning resources and unprecedented acoustic insight. However, an indispensable requisite for singing voice research is the way within which the musical background has being building up to form up a pedagogy for the aforementioned continuum.

Keywords: Cultural Informatics · Pedagogical Agents · Expression in Music

1 Introduction: The Evolution of the Concept of Chromaticism Within Modality

As most calls for "cultural diversity" are coming from the English-speaking world, the world that comes historically from the most westward cultural sphere of influence of the "old world" in the Western hemisphere, it is increasingly pressing, for the sake of clarity, to track the different historical stages in the development of vast enclaves fostering for centuries remarkable forms of culture [1].

Indeed, during the 20th century apparently the organization of doing things in music has acquired an English, or, better say, American orientation for the fundamental principles of pitch organization, for the genres being developed, and, predominantly for the technological acquis that sustains the production and dissemination of soundation activity; it is interesting to see how coexistence may be perceived when taking into account "... civilization ... of the Hindoos, ..., the Saracens, ... (in a certain sense) the Turks" [2].

M. E. Auer et al. (Eds.): IMCL 2025, LNNS 1954, pp. 144–155, 2026.
https://doi.org/10.1007/978-3-032-24998-2_12

Seemingly, all the aforementioned, heterogeneous entities and independent bodies that have shaken the world with the extension of their influence through administration, politics, economics and culture.

Who were the Indo-Europeans, where and when did they appear, what are their ethnic or cultural characteristics would be a thorough study of its own, and certainly, it is beyond the scope of this research [3, 4]. In terms of linguistic spread, the mappings seen in Fig. 1 are rather accurately giving an image of what languages spoken in the 21st century have evolved from the substrate of that continuum [5].

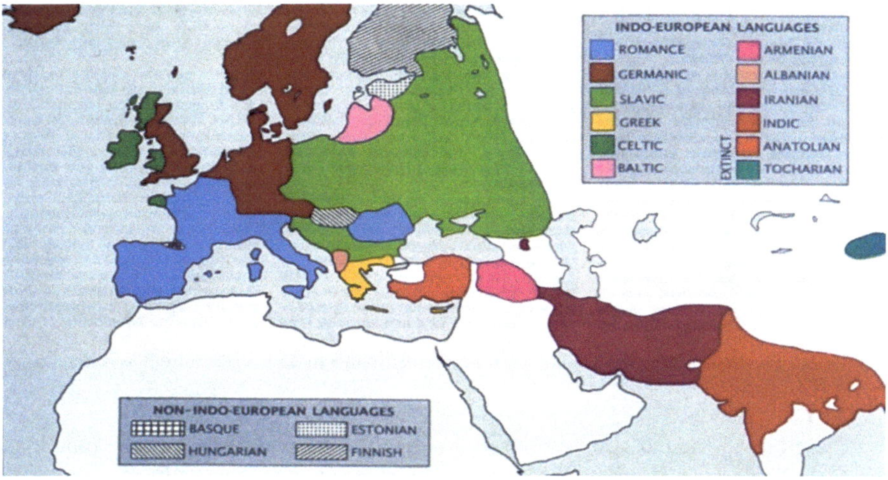

Fig. 1. The 21st century imprint of the Indo-European languages continuum. Source: Wikimedia.

As seen in Fig. 2, the Eastern Roman Empire, branded the last centuries as Byzantium, would be a flourishing continuation of the Roman Empire, in administrative terms, after its division in 395 AD in Eastern and Western parts; however, in cultural values it was a continuation of the Hellenistic civilization, which since the times of Alexander the Great in the 4th century BC had homogenized large regions of South Eastern Europe, Asia Minor, Middle East and Northern Africa by the adoption of Greek customs, and predominantly, the Greek language as *lingua franca* [6]. Nevertheless, cultural processes are mostly two way communications. Factual possessions or immaterial attributes inherited from ancestors morph *per locus* a region's heritage of knowledge [7].

As stated in Encyclopedia Britannica [8], "although Byzantine Music is linked with the spread of Christianity in Greek-speaking areas of the Eastern Roman Empire, it probably derives mostly from Hebrew and early Syrian Christian liturgies (see Syrian chant). Various types of hymns were prominent, among them those called troparion, kontakion, and kanōn".

The key element in correlating Byzantine Ecclesiastical Music with the secular music of the empire is modality. Although the use of the sacred songs from Middle East does not seem compatible with the Greek-Roman cultural continuum, the use of modal melodies and harmonies provides the interconnection, in terms of diachrony and synchrony, with the Hellenization of the Eastern and the Romanization of the Western world Ecumene [from *oikoumenē*, the (inhabited) earth.] [9]. As Landormy puts it, "the music of the Christian Church was derived from the music of the synagogue, modified by the influence of Grecian art." [10].

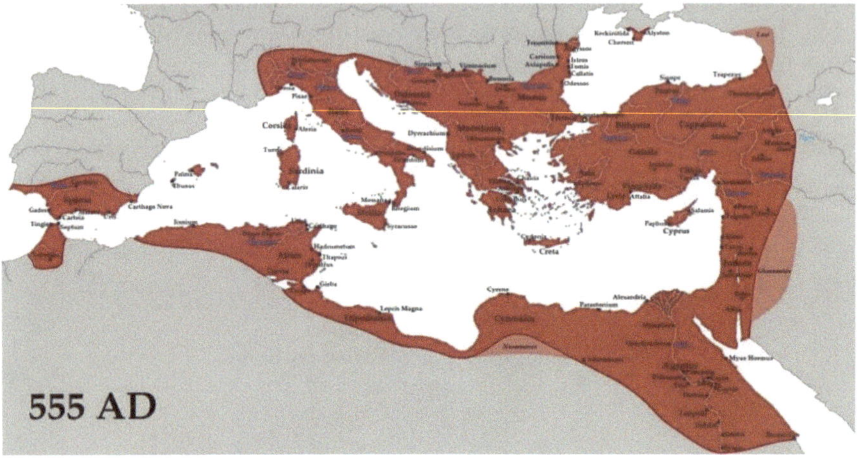

Fig. 2. The Byzantine empire under Emperor Justinian I in 555 AD. Source: Wikimedia.

Therefore, the sound system used in liturgical environments uses broad functional contexts of artistic performance, i.e., music promotes the aesthetic pleasure, amongst others, using schemata from secular music. For instance, Byzantine Ecclesiastical Music is heavily dependent on the concept of "octoechos" – the 8 modes that have evolved from Ancient Greek Music [11].

It should be pointed out that after the times of Cosmas Indicopleustes (6th century, see [9]) the area in perspective, that of the Eastern Mediterranean, witnesses in terms of political and cultural discharge the Arabic flux, and later, the Ottoman spread [12]. As a result, while Indo-European music refers to musical traditions and influences that trace back to the ancient Indo-European-speaking peoples, the last millennium stimuli from Celtic, Germanic, Slavic, and Indo-Iranian groups have further shaped the musical heritage of vast regions in perspective. In any case, if we attempt to provide a comprehensive overview of historical development and cultural significance for the traditional music of the regions seen in Figs. 1 and 2, for the 2nd millennium AD, we will encounter one way or the other the musical system of maqam (or makam in Turkish). It is an alternative way, based on the Arab-Persian way of doing things in music, to delve into the concept of modal music.

Its closest affiliation with the dominant Western Music paradigm comes through the gateways of Ottoman and Turkish musics.

The Arabic word maqām means "place, position, (hierarchical) order", conceptualizing thus the musical systems used, as far as scales are concerned, formulating, the same time, the geographic regions posing influence on the rendering performing characteristic of a certain style. E.g., *hijaz*, from the homonymic Arabic Peninsula, *râst*, from Iran, adopted by Arabic, Turkish and Sephardic Jew musicians, *dügâh* from .... [13]. It is very interesting to point out the diachrony of rearrangements for harmony and discord from the Pythagorean times, as far as scales are concerned, commencing with ancient modes, like *Lydian, Phrygian, Aeolian*, etc. [14], continuing with the Ecclesiastical Modes, transforming then to maqams, and, later on, during the 19th century, transforming the hearing with variants like the *mugam* of Azerbaijan, the *shashmaqam* of Uzbekistan, the *muqam* of the Uyghurs or the *dastgāh* system of Persian music [15].

2 The Research Proposition for Modal Music: Diachrony and Synchronization Diagnoses

The research proposition of this work correlates the music production and distribution in the 'cloud' – a contemporary achievement, with what has been the situation for quite some time [16], when the Indo-European continuum seen in Fig. 1 was disbanded.

Within the digital age music production and dissemination scene, huge amounts of information on the web is exploited to audiophiles.

The abundant flow of information and musical stimuli stretches the scientific context, providing the possibility for linking between the language, society and culture of what the Indo-European continuum has become.

For example, in 1571 the protestant minister Salomon Schweigger (1551–1622) traveled in the Ottoman Empire and transcribed a *mehter havasi*, a military band concert, using a precursor of what would be the Common Music Notation system [17].

Furthermore, indicatively, worthy of attention efforts to bridge the gap in semeiology and perception of music between East-West have been the attemps by:

- Wojciech Bobowski or Ali Ufkî (17th c.). A Polish musician who lived in the Ottoman empire. The system he had devised had similarities with the staff notation that had been establishing itself in Western Europe as the nowadays Common Music Notation scheme – see an exemplar composition in Fig. 3, left.
- Dimitrie Cantemir, of noble Moldavian descent, 18th century. He is known as Kantemiroğlu in Ottoman sources. He has edited a collection of 365 characteristic pieces of the music of his era, known as *Kitâbü İlmi'l-mûsikî alâ vechi'l-hurûfât*. An excerpt of that may be seen in Fig. 3, right.
- Hampartsoum Limondjian (~1820), of Armenian descent, and many others similar.

Taking into account that Thomas Edison launched his *phonograph* in 1877 and Emile Berliner his *gramophone* in 1887, historians and musicologists have a rather hard time in reconstructing music that comes out of diverse notational systems when no recording of it is available [18].

The last two cited musicians, Cantemir and Limondjian, developed neumatic-like semantics, relating to the meaning of whole musical phrases rather than assigning written cyphers representing definite pitch or duration of a single note.

However, although diverse administrative languages, musical notational systems and cultural maneuvers have been traced, when musicologists along with audiophiles commenced to access the original hearing, the exchange of views revealed a very different story in sequels.

Fig. 3. Left: an excerpt from Mecmûa-i Sâz u Söz composed by Ali Ufkî Bey (Bobowski), British Museum, Sloane, nr. 3114, p. 19. Right, an example of Cantemir's notational music Kitâbü İlmi'l-mûsîkî alâ vechi'l-hurufât, İÜ Türkiyat Araştırmaları Enstitüsü Ktp., Hüseyin Sadeddin Arel. Armağanı, nr. Y. 2768, s. 129.

Par example, for the 20th century doings of the Eastern Mediterranean basin, a song for "an Egyptian girl" ("Misirlou" in the Semitic languages of the region) proved a far more straining for the usual attention adventure, standing upon points not easily seen by Westerners from the first glimpse.

Fig. 4. Left: the "American" rendition of Misirlou, as a CD byproduct of a Hollywood success in 1995. Center: The first recording of Misirlou in USA, in vinyl, dating from 1927. Inscribed as "Greek Popular Song". Right: Original Yellow Record Label of a vinyl recorded 21st March,1932, released November 1932 (Matrix No: OK-871–1, Catalogue No: 30–9629, Release Record No: HC 32) with the music of the 1st Cairo Congress of Arab Music.

It is a very different hearing, a trichord "çiftetelli" in D Hijazciar (a maqam resembling is some sense D Minor) performed with 2 *buzukis*, *baglama*, guitar, bass, flute and a bass. The hearing is not rock-pop; rather it sounds like maqams, as those that circulated in a rare recording from the 1st Cairo Congress of Arab Music (« Congrès de Musique Arabe du Caire», discography compiled by Michael Kinnear, Fig. 4, right). In 1932, a remarkable special congress of Arabic musicians was organized by the French musicologist Baron Rodolphe Erlanger with the mentoring of King Faud I of Egypt. This very important congress under the auspices of the National Academy of Music, 22 Malika

Nazly Street (now Ramses Street), Cairo, from the 28 March to the 3 April 1932, gathered dignitaries from the West like Béla Bartók, composer, Paul Hindemith, composer, Alois Haba, composer, Henri Raboud, composer, Emile Vuillermos, music critic, Erich Moritz von Hornbostel [19], musicologist, Robert Lachmann, musicologist, Curt Sachs [19], musicologist, Egon Wellesz, musicologist, Baron Carra de Vaux, scholar, Father Xavier Collangattes, scholar, Henry George Farmer, scholar and musicologist, Alexis Chottin, scholar, Baron Rodolphe d'Erlanger, and a pleiad of Middle East scholars and performers [20].

In organizational terms, it was a deed of prowess. In terms of transfer for musical knowledge it rather was a deed for re-association between East and West, after centuries of segregation. As a matter of fact, when hearing the actual recordings, we more or less make affiliations for many Western musical scales, such as variants of major and minor scales that have have roots in ancient Indo-European traditions. The modal system, used in medieval European music, also traces back to early Indo-European influences.

3 The ITC Agenda: Modifying Subdomains - Intersecting Subdomains

When real hearing is accessible, Machine Learning tools and AI applications may be handy apparatuses for the analyses of sound. Usually, they are cast off as follows:

- Apps for Supervised Learning: by implementing regression-based algorithms, classification, arrangement and cataloguing to well defined datasets (of music).
- Unsupervised Learning: may include algorithms seeking to spot motifs or data structures to loosely circulating data, usually comprising unofficial releases of music.
- Reinforcement Learning: promotes some kind of interaction with an environment or a community, seeking optimization or 'reward' of some kind. Q-Learning or Deep Q-Networks are examples of such agendas.

Seeking, accordingly, to match up descriptions, Reinforcement Learning is somehow vested with NLP, Music Processing and AI apps in the field of ICTs. Representations of Historical Events and their computerized Recreations may be of enormous aid.

For example, old readable manuscripts, like these in Fig. 5, left, are currently available through advanced dissemination portals, as is the Digital Library of the USA Congress. Indeed, it would be a chance of a lifetime encounter till some years ago to access a music parchment, dating from 11th century AD, from the Sinai Peninsula St. Catherine monasterial library, Egypt [21].

Some distinguished scholars from the Balkans have been recreating the song itself [22, 23], in terms of musicality, while in another project, more USA-centric in specific, they have managed to procreate the acoustics for the hearing, with exactness in description, by rendering virtual reality simulations [24]. On another occasion, more recently, they described the process under which the formative potential for high-level singing was standardized in 12th century imperial courts [25].

Even further, more recent digitalization processes have made available to scholars and the public similar parchments from Holy Mountain, Greece. For instance, in Fig. 5,

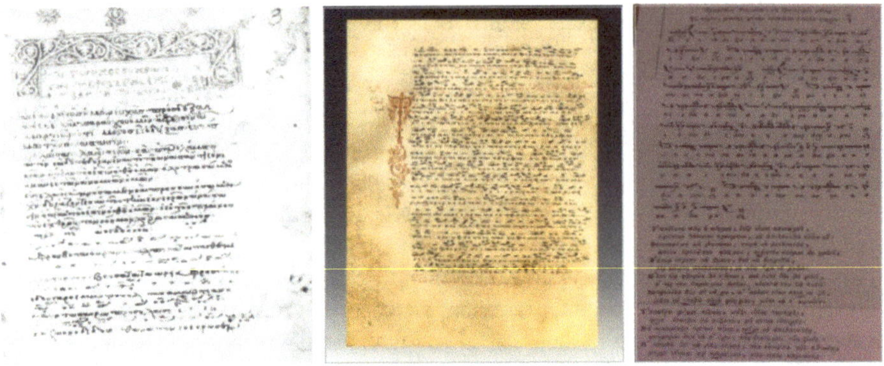

Fig. 5. Left: Oktoëchos. Greek Manuscripts 795. 1100 AD. Manuscript/Mixed Material. Digital Library of the USA Congress https://www.loc.gov/item/00271074414-ms/. Center: Vatopedi Monastery Digital Library Collection manuscript # 1486 (p. 192b), 1292 AD. Right: A "Hisar Puselik" secular song in Mode 5, published in "Euterpe", Constantinople (Istanbul) 1830 AD.

centre, one may see a *Sticherarion* of the 13th century. It is manuscript # 1486 (p. 192b) which dates from 1292 AD. It is part of the Vatopedi Monastery collection [26].

Last, but not least in this sequel, is a digital imprint of a 19th century folk-pop song of Mode 5 (= Mode Plagal A′). It was published in a music book titled "Euterpe" (after the homonymic muse) in Constantinople (Istanbul) at 1830 [27]. It was a collection of secular songs.

The performing style is described as "Plagal A', hard, diatonic, spatheios (σπάθη, spathi = "sword", also the name of a scale alteration sign) as far as its mode, gender and accentuations are concerned; it is described as a secular chant, prescribing thus a rather lascivious rendition. Connected with the maqam rendition specifics, in a hint of "Supervised Learning" intersection, it is suggested as "Hisar Puselik" style.

This song had also been published in *Pandora*, some 50 years later [28] while it has been spotted, as a variant, in codex #322 of the Docheiariou Monastery [29].

The goal, in any case, is seeking to find out the complex link between various variables and shed light on pathways to promote a culture of integrating AI into Humanities and Social Sciences (HSS) education. Something like the research that conducted Idelsohn in 1914 for Jewish and Arabic Music [30].

Through tackling the complexities of AI adoption and clarifying the implications for learning, this research endeavors to provide practical insights that can help scholars embrace AI tools for inclusivity in the field of musicality disseminated through cybercultures [31].

Again, as there is absence of sound files for melodies that have survived from a world centuries ago, the only way, perhaps, to pick up the trail for reconstructing the actual hearing is to decipher the traits of "music theory" of that time. Usually, scholars are well acquainted with what could be called "the grammar of Western music" in "the field that studies how music works."

It is not that easy, however, when the (musical) "linguistics" or the "secrets of the trade" are encountered, for musics that have flourished centuries ago within the Indo-European continuum.

Some considerable progress may be seen the last few years when musics from that era survive in Byzantine Music style notation [1].

Through addressing the opportunities and challenges associated with AI integration, this study provides a handy roadmap for scholars, musicians, educators, and the general public to seek, to navigate and to examine the complexities of the digital age music dissemination, as far as the Indo-European continuum is involved.

Historically, the Eastern Mediterranean basin is linked to such doings for more than 3000 recorded (in writing) events. Even further, highly prolific proceedings of modern musicality have linked people and places with dominant Western countries, over both the Mediterranean and the Atlantic as well. Therefore, what rests to be seen is how technological progress may be engulfed in a cultural campaign over the 21st century's Indo-European route (Fig. 1).

According to AI agents, like Copilot™: "Indo-European musical traditions have had a profound impact on contemporary music, shaping genres, instruments, and even musical theory. Here are some key influences:

Scales and Modes: Many Western musical scales, such as the major and minor scales, have roots in ancient Indo-European traditions. The modal system used in medieval European music also traces back to early Indo-European influences.
Instruments: String instruments like the lute, which evolved into the modern guitar, have Indo-European origins. The violin, harp, and bagpipes also have deep connections to Indo-European cultures.
Rhythmic Patterns: Folk traditions from Indo-European regions, such as Celtic, Slavic, and Indo-Iranian music, have contributed unique rhythmic structures to contemporary genres like rock, jazz, and electronic music.
Cross-Cultural Fusion: Indo-European musical elements have blended with other traditions, leading to hybrid genres. For example, Indian classical music has influenced Western rock and pop, while Slavic folk music has shaped electronic and experimental sounds." [34]

4 ICT and AI Support: Cognitive Walk-Through Within the Indo-European Continuum

Under normal circumstances, some years ago, the topic of this research would be well versed for certain scholars of Music; it would be not discussed, though, further than that. However, recent advances in ICTs have raised this troubling issue within the sphere of Digital Pedagogy (for the domain of Education) and resilient adaptation (for the business domain of entertainment).

In practice, there is a synergy between humans (i.e., the actual learners, the social actors, and, finally, the end-consumers) and AI agents (i.e., the facilitators, the "middleware" search engine mechanisms, and, ending up, the generating and enabling content technologies) that occasionally produce stronger results than what classic, "individual" players (like Hollywood or Bollywood industrialists) would have envisioned.

Therefore, users of commercial digital stores (like Apple Music™, Spotify™, …) or ICTs enabling streaming (like YouTube™, Tik-Tok™, …) organize media into custom albums and playlists widely available with little cost.

A very interesting point, for the advantage of revitalized interest in what would be the music of the 21st century Indo-European continuum, comes from the computer industry: mostly the physical components (hardware, mobile devices, sensors, etc.) are made in Asia, while the software (as stand-alone apps or web-accessed services) lies in Western World (American, European) hubs and cloud services. This readily anticipated belief seems not to be the case for the near future, as a trend-line, some findings of this research have been pointing to.

A research was conducted within the Aristotle University of Thessaloniki in June 2025. Some 30 technology students (21 male, 9 female) participated in this research, most of them coming from the School of Informatics, while a few were from the School of Electrical and Computer Engineering. They were focusing on Mobile Device Interfaces.

They were asked to contribute in morphing the profile of a hi-tech user of media services, as far as music entertainment is concerned. Albeit rather advanced for the time being, however, it is expected to be the norm in the very next few years to come for the average user.

The inquiry processed in three steps, revealing more precisely how attention for a specific genre or style of music may lead to Reinforced creative Learning.

First Step:

The users were asked if they access music via their mobile devices. Indeed, they all affirmed that primarily their smartphone is the source for music entertainment, although their personal computers or laptops were occasionally used for that as well.

Second Step:

The users were not asked deliberately to preamble broadband TV services for this step, as there is a plausible constraint in most countries on what a viewer may access. For example, in Greece there is limited access to channels originated from Asia, Africa, America or even other European countries.

Furthermore, while not everybody may have a laptop, PC, Radio or TV set for accessing music, it is expected to have a smart mobile device committed so. Indeed, via it all the participants confirmed having access to music in the domain of thousands of songs, either via subscription services with global vendors like Spotify™ or by other streaming services, like YouTube™, which are currently free of charge or regional restrictions.

The situation described in the First Step seems to be compatible with what digital broadcasting norms have constrained for audiovisual services within a country. It would be expected for an average user of public-access television to be restricted in having admittance to the cultural doings of a country in a Far East region or even a Middle East - North African location.

Nevertheless, the participants in this research had surprisingly a more global orientation as far as their music proliferation emerged.

In Fig. 6 can be seen what kind of music they have accessed in the first semester of 2025.

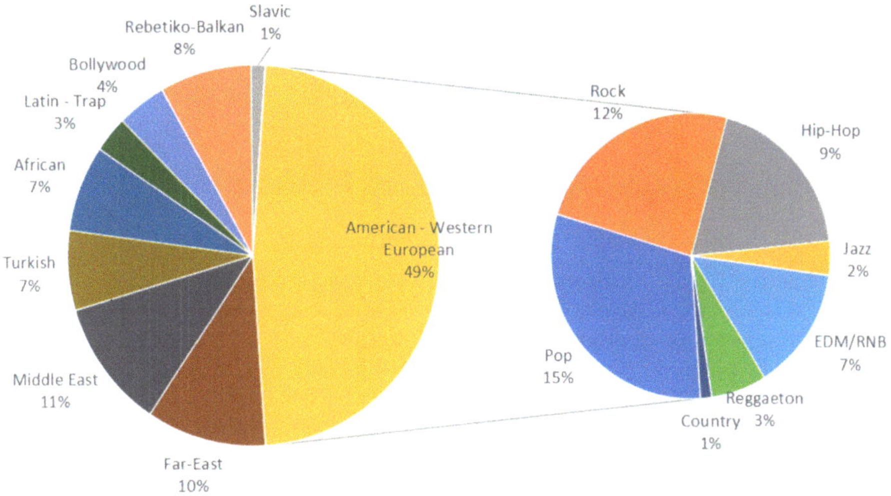

Fig. 6. The genres of music the participants of the survey accessed via ICT networks in the first semester of 2025.

Surprisingly, apart from the dominant American & Western European music, they have started accessing other genres that would be characterized in circumstances different from those present as "exotic". Technology plays a crucial role in fostering the development of an ecosystem that promotes and sustains such preferences. For instance, the interest for "Far Eastern" music does not have a geographical affinity but, rather, a technological one. Users, apart from notable artifacts coming from Japan or Korea entangling the investigation procedures for their use, are also attracted to new styles like J-pop and K-pop. In the same manner that music comes out of Hollywood, as a byproduct, it flows from Bollywood, making it a noteworthy contributor. Also, to have some local intricate affiliations with Balkan or Turkish music sounds rather expected, having "Misirlou" in mind, but this attitude for modal music is enhanced by extending this interest to Middle East and North African cultural impressions as well.

Third Step:

The situation described in the Second Step, as visualized in Fig. 6, is forming a hypothesis about cultural bias in conjecture with the technological acquis.

To some extent, it seems that AI, advanced Search Engines or Machine Learning Tools capture the interest of users by exciting their curiosity to deploy content creators for multimedia productions. Recently, quite a few such tools were made freely available for "public activism", in a way similar to social networks functioning as tools for social behaviorism or "political activism". Notable, amongst others, are: AIVA™, Beatoven™, Mubert™, SoundRaw™, …, etc.

The participants of this research were asked whether they have used any of these tools, with emphasis paid to perhaps the most prominent of those, SoundRaw™, for better interaction with their music ecosystem.

Fig. 7. Left, and Center: the SoundRaw™ cloud-based platform for accessing and interacting with global music with the help of AI tools and Machine Learning apps. Right: the rate in which users of this survey were attracted to this computer structure.

As seen in Fig. 7, some 43% of them (13/30) were already attracted and had somehow used SoundRaw™ (or similar) tools for recreating the music they had been interacting with.

It is noteworthy that these tools do not have as previous stipulation the knowledge of music theory, part of which was presented in Sects. 2 & 3 of this paper, to further proceed with the rapid prototyping of modal music or the algorithmic composition enhancements offered by the AI support substrate.

References

1. Griffiths, P., Griffiths, P.: The New Penguin Dictionary of Music. Penguin Books, London (2006)
2. Pelican, J.: The Idea of the University - A Reexamination. Yale University Press (1992)
3. Kallio, P., Koivulehto, J.: More remote relationships of Proto-Indo-European. In: Klein, J., Joseph, B., Fritz, M., (eds.) Handbook of Comparative and Historical Indo-European Linguistics, PP. 2280–2291 (2018)
4. Renfrew, C.: Archaeology & Language. Jonathan Cape, The Puzzle of the Indo-European Origins. London (1987)
5. Hugoe, M.P.: The concise Oxford dictionary of linguistics, 3rd ed. Oxford University Press, Oxford (2014)
6. Droysen, J.: Geschichte des Hellenismus, Vols. I-III, F.A. Berthes, Gotha (1877)
7. Dutta, B., Jackson, M.O.: On the Formation of Networks and Groups. In: Dutta, B., Jackson, M.O. (eds.) Networks and Groups: Models of Strategic Formation. Springer-Verlag, Heidelberg (2003)
8. Encyclopedia Brittanica, Online Edition (Accessed: June 2025). http://britannica.com
9. McCrindle, J.W.: The Christian Topography of Cosmas, an Egyptian Monk, pp. 122–23, 1897, rpt. New York: Burt Franklin (1964)
10. Landormy, P.: A History of Music, London-New York: Charles Scribner's Sons (1923). English translation of Histoire de la musique, Paris (1910)
11. Barsky, V.: Chromaticism. Kluwer, The Netherlands (1996)
12. Signell, K.: Makam. Da Capo Press, Incorporated (1986)
13. Zanos, I.: Ichos und Makam, vergleichende Untersuchungen zum Tonsystem der griechisch orthodoxen Kirchenmusik und der türkischen Kunstmusik. Orpheus-Verlag, Bonn (1994)
14. Jeans, J.: Science & Music. Cambridge University Press, Cambridge (1953)

15. Farraj, J., Abu Shumays, S.: Inside Arabic Music: Arabic Maqam Performance and Theory in the 20th Century. Oxford University Press (2019)
16. Jackson, M.O., Yariv, L.: Diffusion on Social Networks. Économie Publique **16**(1), 3–16 (2005)
17. Ben-Zaken, A.: Cross-Cultural Scientific Exchanges in the Eastern Mediterranean, PP. 1560–1660. JHU Press (2010)
18. Politis, D.: Γλώσσες & Διεπαφές στη Μουσική Πληροφορική. Klidarithmos Publications, Athens (2007)
19. Congrès de Musique Arabe du Caire, George Walter Dillnutt recording and producing Original Yellow Record Label Recorded 21st March,1932, Released November 1932, Matrix No: OK-871-1, Catalogue No: 30-9629, Release Record No: HC 32
20. Greek Manuscripts 795. Oktoëchos Oktoechos. Greek Manuscripts 795. Oktoëchos Oktoechos. 1100. Manuscript/Mixed Material. https://www.loc.gov/item/00271074414-ms/
21. Pelican, J.: The Idea of the University - A Reexamination. Yale University Press (1992)
22. Pentcheva, B.: Hagia Sophia: Sound, Space, and Spirit in Byzantium. Penn State University Press (2017)
23. Capella Romana – Lingas, A.: Lost Voices of Hagia Sophia: Medieval Byzantine Chant, CD, EAN/UPC: 746839858954, Catalog No: CR-420 (2019)
24. Alexandru, M.: Paleography of Byzantine Music. Open digital resources on educational technology (2017). https://repository.kallipos.gr/handle/11419/6487
25. Spyrakou, E.: The Byzantine choral system until 1204. Byzantine Musical Culture. Papers of the 1st International Conference of the American Society of Byzantine Music and Hymnology, Attica, pp. 144–156 (2007)
26. Dumbarton Oaks, Online Exhibits & Projects 1703 32nd Street NW Washington, DC 20007 https://www.doaks.org/resources/all
27. Euterpe. Χουρμουζίου Χαρτοφύλακος, Galata Printing House, Istanbul (1830)
28. Pandora (Πανδώρα) – periodical edition 1849–1880. «Μουσικόν Απάνθισμα» (1873), https://el.wikisource.org/wiki/
29. https://en.wikipedia.org/wiki/Docheiariou
30. Idelsohn, A.Z.: Hebräisch-Orientalischer Melodienschatz. I. Band: Gesänge der Jemenischen Juden. Zum ersten Male gesammelt, erläutert und herausgegeben von A.Z. Idelsohn, Leipzig: Breitkopf und Härtel, pp. xi+158 (1914)
31. Bell, D.: An Introduction to Cybercultures. Routlege (2002)

Conversational AI

Growing Adoption of Conversational Agents for Workplace Learning: A Scoping Review

Eirini Voutsa(✉) and Stavros Demetriadis

School of Informatics, Aristotle University of Thessaloniki, Thessaloniki, Greece
{eirinivoutsa,sdemetri}@csd.auth.gr

Abstract. While conversational agents can streamline corporate learning, technological and ethical concerns challenge their adoption. Overcoming these obstacles is crucial to realize their professional development benefits. The goal of this study is to identify the role and impact of conversational agents in corporate environments and to investigate the essential conditions for their utilization to advance corporate learning. To conduct this study, the PRISMA model and keyword-based search were employed to explore scientific articles published between 2020–2024 on the characteristics and organizational utilization of conversational agents, selecting those that referenced or highlighted opportunities for workplace learning. Based on our review, the seamless integration of conversational agents into workplace learning requires commitment from both organizations and employees: Organizations should drive effective implementation by ensuring resources and transparency, while employees should actively use the agents. This dual approach is necessary to create agents that reflect corporate culture and meet employee needs. Our findings indicate the need for the adoption of proper conversational agents design within corporate frameworks and the expansion of employee awareness regarding the agents' capabilities, limitations, and appropriate use.

Keywords: Corporate Training · Conversational Agents · Workplace Learning

1 Introduction

1.1 Background

Conversational Agents (CAs) are Artificial Intelligence (AI) tools that can bring significant value to any organization, whether they supplement the work of employees, or handle specific tasks completely [1]. Their main purpose is to provide instant and accurate information on demand [2]. However, their application goes beyond simple information access; they can be powerful tools for workplace learning and development [3]. Specifically, by integrating CAs into the daily workflow, employees are empowered to train on new aspects of their job and to directly seek specialized information relevant to their corporate environment [4]. At the same time, by analyzing user interactions, organizations can gain critical insights into employees' concerns, understanding deeper their needs and how they perform in their roles. This integration results in a mutually

M. E. Auer et al. (Eds.): IMCL 2025, LNNS 1954, pp. 159–166, 2026.
https://doi.org/10.1007/978-3-032-24998-2_13

beneficial outcome. Employees acquire valuable assets for their professional growth and receive efficiently the corporate information they need, while organizations can make data-driven decisions aligned with the real needs and capabilities of their workforce [5].

1.2 Objectives

This scoping review aims to identify the role and impact of CAs on the most common business domains and processes and explore the essential conditions for their utilization to advance workplace learning. The following research questions are posed.

RQ 1: What are the main business domains leveraging CAs and how do these agents bring workplace learning opportunities to these domains?

RQ 2: What are the benefits of CAs usage for employees and organizations?

RQ 3: What are the concerns of CAs adoption?

2 Methods

2.1 Search Strategy and Eligibility Criteria

A scoping review was conducted aligned with PRISMA-ScR (Preferred Reporting Items for Systematic Reviews and Meta-Analyses Extension for Scoping Reviews) guidelines [6]. For this review, a literature search was performed between November 2024 and January 2025 using the Scopus database and Google Scholar. For the Google Scholar search outcomes, only the first 5 pages of results (equivalent to 50 citations) for each search query were screened, because the relevance of citations was observed to decrease significantly thereafter.

The search strategy in the Scopus database combined terms for conversational agents such as "conversational ai", "chatbots", "chatGPT" and keywords such as "employee", "workforce", "workplace learning", "corporate training" and "employee training", using Boolean operators (AND, OR). In Google Scholar, broader keyword phrases like "Conversational Agents in Workplace" were also used to ensure a more comprehensive range of results.

The search focused on studies published in English between 2020–2024. Only studies with freely accessible full text were considered for review.

2.2 Screening, Data Extraction and Synthesis

A two-stage screening process was used for study selection, starting with a title and abstract review, followed by full-text analysis. Studies were excluded if they focused on general AI rather than Conversational AI or lacked direct relevance to the research questions. Using a standardized form, key data—including author, year, CA type, and findings—were extracted. These data were then narratively synthesized into three overarching themes: business domains, benefits, and concerns.

3 Results

3.1 Study Selection and Characteristics of the Included Studies

From an initial 374 retrieved studies, 53 remained after the removal of duplicates and initial screening. A final 26 studies were included after full-text review and are presented in Table 1.

Table 1. Distribution of Included Studies per Year

Year of Study	Study Reference	Number of Studies
2020	[7]	1
2021	[8–11]	4
2022	[3, 4, 12]	3
2023	[2, 13–20]	9
2024	[5, 21–28]	9

3.2 RQ 1: What Are the Main Business Domains Leveraging CAs and How Do These Agents Bring Workplace Learning Opportunities to These Domains?

Human Resources (HR). CAs can effectively handle a range of HR tasks. In the pre-hiring phase, a CA acts as a Recruiter, managing talent acquisition [2, 13, 21] by screening candidates and conducting interviews to gain deeper understanding into their overall fit for the job role [2, 8, 13]. Drawing on its AI capabilities, CA streamlines the recruitment process, trying to apply objective criteria to evaluate candidates, ensuring an unbiased selection process [2]. During the hiring phase, a CA can operate as an Onboarding Assistant, providing introductory information and responding to inquiries from newly hired employees to help them adjust to the corporate culture and familiarize themselves with the new working environment [2, 5, 8, 14, 22]. In the post-hiring phase, a CA can serve as an HR Support Agent, answering FAQs and addressing incoming requests [8, 12, 14, 21].

Operations and Employee Support. CAs can serve as valuable support tools for employees by fulfilling three distinct roles: assistants, Colleagues, and Mentors. As Assistants, they help employees to handle routine tasks [7, 9, 10, 13, 15, 16, 22] such as managing emails [10, 11, 22, 23, 27] and translating texts [13, 14, 17, 18]. They also provide quick access to information by efficiently retrieving data from extensive databases [2, 4, 5, 8, 22–24]. As Colleagues, they collaborate with employees to answer their queries and assist them in garbled issues [4, 5, 11, 13, 15]. As Mentors, they provide guidance in problem-solving and facilitate knowledge acquisition. This involves not only delivering information but also offering tailored explanations and insights. Moreover, CAs as Mentors can generate new knowledge which contributes to the resolution of operational difficulties [5, 8, 13, 18, 19, 21, 22, 24, 27].

Employee Training. CAs can play a crucial role in corporate training activities. They are capable of transferring knowledge clearly and efficiently, answering queries, and clarifying complex concepts [2, 8]. CAs also have the potential to recommend appropriate training options and create customized learning paths that help employees acquire essential skills relevant to their professional roles [8, 26]. Additionally, CAs can perform various learning tasks, enabling employees to engage in an interactive and guided learning experience [2, 3, 21, 22].

Assessment Processes. CAs can undertake assessment processes such as employee performance evaluation [2, 8]. They can also estimate the mental health and the emotional state of employees based on their interactions, identifying situations such as demotivation, burnout, disappointment, and distress. In case CAs identify an issue, they can be trained to support employees accordingly [2, 8, 10, 17, 21].

Feedback Collection. CAs can be a feedback collection tool [2, 8], conducting polls and surveys [10, 13, 21]. The data collected from these interactions can be analyzed to provide new knowledge and real-time insights, which help organizations improve their decision-making mindset [2, 5, 13, 15, 19].

3.3 RQ 2: What Are the Benefits of CAs Usage for Employees and Organizations?

Benefits for Employees. Employees can benefit from CAs usage in their daily work routine. CAs enable communication in natural language, without requiring technical or specialized knowledge to use them [2, 5, 7, 9, 13, 17, 24]. Their interaction method is inherently friendly and intuitive [4, 5, 11, 22, 24], since they function much like popular chat platforms employees are already accustomed to [8, 27]. CAs are capable of providing immediate [2, 8, 10, 12, 20, 22], around the clock responses independent of conventional business hours [14–17, 21, 23]. They are accessible from multiple devices allowing employees to interact from anywhere [8, 14].

CAs can also free up employees from certain tasks and help them manage large volumes of open issues, which in turn reduces their workload [8, 9, 13–15, 17, 23]. They can also improve employee performance by providing access to multiple data sources and delivering instant, complex, or combined information [5, 7, 9, 13–15, 17, 21, 23, 24]. Consequently, CAs support faster decision-making and problem-solving [5, 8, 13, 21–24]. Moreover, employees can use CAs to search for and acquire new knowledge or even discover new ideas [15].

CAs can provide personalized communication experience, tailored to employees' needs [2–4, 10, 13, 16–19, 21, 22, 24]. Through interactions with employees, CAs can detect their emotional state and guide the dialogue, which fosters employee motivation and engagement [10, 21, 24, 26]. Furthermore, the absence of a human conversational partner can encourage employees to express their thoughts freely, minimizing any stress they might feel about being judged for their queries [8].

Benefits for Organizations. The utilization of CAs by organizations can automate standardized processes [7, 8, 16, 18, 19, 25], increase employee productivity [2, 4, 9, 13–17, 21, 22, 27], reduce the man-day effort required for certain tasks [5, 8, 18, 24], and contribute to cost savings [3, 4, 8, 9, 13, 16, 21, 24, 27].

CAs are continuously available and can serve multiple employees simultaneously [16, 22]. Their capability of prompt responsiveness and effective sharing of information or new knowledge can enhance employee engagement in the corporate setting [2, 5, 13, 21]. In addition, CAs can provide valuable insights based on their interactions with employees to advance corporate decision-making [2, 5, 8, 13, 19].

3.4 RQ 3: What Are the Concerns of CAs Adoption?

Resource Investment. A considerable investment in time is necessary to ensure a CA operates efficiently, to continuously improve its functionality, and to familiarize employees with its capabilities and effective utilization. During the early phases of CA development, organizations should anticipate substantial costs and potential complexities, particularly concerning the integration with existing back-office corporate systems [2, 8, 13, 22].

The Black Box Problem. Once integrated and made available, CAs often face skepticism when compared to traditional tools and methods that handle similar tasks. A significant factor contributing to this is the lack of transparency about how CAs operate, which often remains a black-box for users. Thus, employees may find it difficult to trust CAs with their personal data [2, 3, 12–15, 18, 24, 25].

Training Data Bias. An essential aspect of CAs powered by AI models is the validity of their responses, which can be compromised by inherent biases within the training data. This may result in outputs that contain incorrect information or inappropriate material. When CAs responses fail to meet employee expectations, it can cause disappointment and frustration [2, 8, 13, 16, 18, 22, 27, 28].

New Job Roles. The adoption of CAs brings about changes in job roles, as machines increasingly undertake tasks previously performed by humans [2, 13, 18]. An increased dependency of corporate processes on CAs will likely strengthen human-machine interaction, which could reduce interpersonal professional relationships [22].

Inequalities. Inequalities may occur between employees who possess strong skills with AI technologies like CAs and those less familiar with them. Similar disparities may also emerge between large organizations, which can allocate more resources to implement and maintain advanced CAs, and smaller ones that cannot [13, 18].

4 Discussion

4.1 Principal Findings

The scoping review indicates that CAs create and support workplace learning opportunities. For employees, they act as continuous learning tools that offer on-demand support for new tasks, guide problem-solving, and deliver personalized communication. For organizations, CAs contribute to corporate learning by standardizing processes and by providing insights derived from employee interactions. This approach effectively embeds learning directly into the daily workflow. However, literature also identifies significant challenges to CAs adoption, which exist at technical and ethical levels. Technical challenges include the substantial resource investment required for their development, integration, and maintenance. Key ethical concerns are the lack of transparency in how CAs operate, which can reduce employee trust, and the potential for biased training data to produce inaccurate or unfair outcomes. Together, these issues lead to socio-organizational risks, such as undermining the intended learning benefits and potentially widening the skill gap between employees.

4.2 Implications

Growing adoption of CAs into the workplace should begin with a transparent and strategic framework [18]. From the outset, it is necessary for organizations to clearly define their requirements, prepare training data to align with internal policies [5], communicate the technology's purpose, and specify the tasks the CA can reliably perform to effectively manage employee expectations [9, 20]. CAs integration into the workplace must be guided by a human-centric methodology, aiming to augment the workforce rather than replace it [2]. Design principles, therefore, need to prioritize personalized interactions that facilitate harmonious collaboration between employees and CAs [3, 11, 18]. CAs should also be built to learn from these interactions, creating a feedback loop for their continuous improvement [3, 5, 20]. For employees to embrace CAs, trust is essential [25]. This means encouraging hands-on exploration of the benefits of CAs and always providing an option for escalation to human support [20]. Finally, successful adoption of CAs requires a dedicated approach to workforce development. Through reskilling and upskilling initiatives, organizations can empower their employees to adapt. Targeted training is critical to ensure everyone shares a common understanding of how to use CAs effectively to achieve corporate goals [13, 18, 23].

4.3 Conclusions

This scoping review has identified the main business domains where CAs are utilized for workplace learning. The findings indicate that CAs are capable of executing various tasks, supporting employees and streamlining corporate processes. The review also highlights the significant concerns that arise from the integration of CAs into corporate environments. It is evident that a dual approach is necessary to create CAs that both reflect the corporate culture and meet employee needs. On one hand, organizations should allocate resources for effective implementation, follow design principles that align with their culture, ensure transparency at ethical and technological levels, and promote awareness initiatives. On the other hand, employees should be motivated to explore the functionalities of CAs and discover how to benefit from their use. The adoption of CAs is a shared responsibility for both organizations and employees. When this responsibility is successfully shared, CAs can become an asset in the workplace.

References

1. Felten, E., Raj, M., Seamans, R.: How will Language Modelers like ChatGPT Affect Occupations and Industries? (2023). http://arxiv.org/abs/2303.01157

2. Rane, N.: Role and Challenges of ChatGPT and Similar Generative Artificial Intelligence in Human Resource Management. SSRN Journal (2023)
3. Feng, S.: Designing Chatbots for Workplace Learning. PACIS 2022 Proceedings (2022)
4. Gkinko, L., Elbanna, A.: Hope, tolerance and empathy: employees' emotions when using an AI-enabled chatbot in a digitalised workplace. ITP. **35**, 1714–1743 (2022)
5. Sumbal, M.S., Amber, Q.: ChatGPT: a game changer for knowledge management in organizations. K. **54**, 3217–3237 (2024)
6. Tricco, A.C., et al.: PRISMA extension for scoping reviews (PRISMA-ScR): checklist and explanation. Ann. Intern. Med. **169**, 467–473 (2018)
7. Meyer von Wolff, R., Hobert, S., Masuch, K., Schumann, M.: Chatbots at Digital Workplaces – A Grounded-Theory Approach for Surveying Application Areas and Objectives. Pacific Asia J. Ass. Info. Sys. 12 (2020)
8. Majumder, S., Mondal, A.: Are chatbots really useful for human resource management? Int. J. Speech Technol. **24**, 969–977 (2021)
9. Brachten, F., Kissmer, T., Stieglitz, S.: The acceptance of chatbots in an enterprise context – A survey study. Int. J. Inf. Manage. **60**, 102375 (2021)
10. Hungerbuehler, I., Daley, K., Cavanagh, K., Garcia Claro, H., Kapps, M.: Chatbot-Based Assessment of Employees' Mental Health: Design Process and Pilot Implementation. JMIR Form Res. **5**, e21678 (2021)
11. Sowa, K., Przegalinska, A., Ciechanowski, L.: Cobots in knowledge work: Human – AI collaboration in managerial professions. J. Bus. Res. **125**, 135–142 (2021)
12. Charlwood, A., Guenole, N.: Can HR adapt to the paradoxes of artificial intelligence? Hum. Resour. Manag. J. **32**, 729–742 (2022)
13. George, A.S., Hovan George, A.S., Gabrio Martin, A.S.: ChatGPT and the Future of Work: A Comprehensive Analysis of AI's Impact on Jobs and Employment (2023)
14. Gkinko, L., Elbanna, A.: Designing trust: The formation of employees' trust in conversational AI in the digital workplace. J. Bus. Res. **158**, 113707 (2023)
15. Wang, X., Lin, X., Shao, B.: Artificial intelligence changes the way we work: A close look at innovating with chatbots. J. Am. Soc. Inf. Sci. **74**, 339–353 (2023)
16. Raj, R., Singh, A., Kumar, V., Verma, P.: Analyzing the potential benefits and use cases of ChatGPT as a tool for improving the efficiency and effectiveness of business operations. BenchCouncil Transactions on Benchmarks, Standards and Evaluations **3**, 100140 (2023)
17. Gkinko, L., Elbanna, A.: The appropriation of conversational AI in the workplace: a taxonomy of AI chatbot users. Int. J. Inf. Manage. **69**, 102568 (2023)
18. Fui-Hoon Nah, F., Zheng, R., Cai, J., Siau, K., Chen, L.: Generative AI and ChatGPT: Applications, challenges, and AI-human collaboration. J. Info. Technol. Case and Appl. Res. **25**, 277–304 (2023)
19. Rane, N.: ChatGPT and Similar Generative Artificial Intelligence (AI) for Smart Industry: Role, Challenges and Opportunities for Industry 4.0, Industry 5.0 and Society 5.0 (2023). https://papers.ssrn.com/abstract=4603234
20. Kecht, C., Egger, A., Kratsch, W., Roeglinger, M.: Quantifying chatbots' ability to learn business processes. Information Systems, 102176 (2023)
21. Pillai, R., Ghanghorkar, Y., Sivathanu, B., Algharabat, R., Rana, N.P.: Adoption of artificial intelligence (AI) based employee experience (EEX) chatbots. ITP. **37**, 449–478 (2024)
22. Valtolina, S.: Designing Conversational Agents for Empowering Human Work. Presented at the CoPDA@AVI (2024)
23. Retkowsky, J., Hafermalz, E., Huysman, M.: Managing a ChatGPT-empowered workforce: Understanding its affordances and side effects. Bus. Horiz. **67**, 511–523 (2024)
24. Fahad, S.A., Salloum, S.A., Shaalan, K.: The Role of ChatGpt in Knowledge Sharing and Collaboration Within Digital Workplaces: A Systematic Review. In: Al-Marzouqi, A., Salloum,

S.A., Al-Saidat, M., Aburayya, A., Gupta, B. (eds.) Artificial Intelligence in Education: The Power and Dangers of ChatGPT in the Classroom, pp. 259–282. Springer Nature Switzerland, Cham (2024)

25. Ramaul, L., Ritala, P., Ruokonen, M.: Creational and conversational AI affordances: How the new breed of chatbots are revolutionizing the knowledge industries. Business Horizons, 67 (2024)
26. Tusquellas, N., Palau, R., Santiago, R.: Analysis of the potential of artificial intelligence for professional development and talent management: A systematic literature review. Int. J. Info. Manage. Data Insights **4**, 100288 (2024)
27. Ritala, P., Ruokonen, M., Ramaul, L.: Transforming boundaries: how does ChatGPT change knowledge work? JBS. **45**, 214–220 (2024)
28. Hannigan, T.R., McCarthy, I.P., Spicer, A.: Beware of botshit: How to manage the epistemic risks of generative chatbots. Bus. Horiz. **67**, 471–486 (2024)

Early Insights from STI Shield: A Multi-chatbot Mobile Solution for Sexual Health Education in African Young Adults

Hemad Fetrati(✉), Gerry Chan, and Rita Orji

Faculty of Computer Science, Dalhousie University, Halifax, NS, Canada
hfetrati@dal.ca

Abstract. Sub-Saharan Africa bears the highest burden of sexually transmitted infections (STIs) globally. Young adults are particularly at risk due to limited access to sexual health education and high prevalence of risky behaviors. STI Shield is a mobile app developed to address these challenges by delivering sexual health education through a hybrid approach that combines rule-based and large language model chatbots. Grounded in the Health Belief Model and persuasive system design principles, the app provides structured learning and adaptive counseling to promote safer sexual practices and informed decision-making. Early results from a mixed-methods evaluation showed improved STI knowledge, strong perceived persuasiveness, and high usability. These findings suggest that STI Shield is a user-friendly, engaging, and effective tool for enhancing sexual health literacy and supporting behavior change among African young adults. The study contributes to a growing understanding of how conversational AI can help close critical health education gaps in underserved populations.

Keywords: African young adults · chatbot · conversational agent · gamification · large language model (LLM) · mobile health (mHealth) · serious game · sexually transmitted infection (STI)

1 Introduction

Sexually transmitted infections (STIs) remain a major public health concern in sub-Saharan Africa. Young adults are especially vulnerable due to higher prevalence of risky sexual behaviors and limited access to sexual health education [1]. According to the United Nations, only 37% of young people in sub-Saharan Africa have adequate knowledge about HIV prevention and transmission, underscoring a persistent gap in education and awareness in this region [2].

With smartphone ownership in sub-Saharan Africa projected to reach 65% by the end of 2025 [3], mobile health (mHealth) offers a powerful opportunity to address sexual health challenges. Defined as the use of smartphone and tablet

M. E. Auer et al. (Eds.): IMCL 2025, LNNS 1954, pp. 167–175, 2026.
https://doi.org/10.1007/978-3-032-24998-2_14

apps to support health and wellness [4], mHealth interventions have shown success in increasing uptake of sexual and reproductive health services among young Africans [5]. However, there remains a need for theoretically informed and culturally tailored interventions that incorporate effective behavior change techniques [5], as designing content that reflects users' social and cultural realities has been shown to be more effective than generic approaches [6]. This is particularly important for the African young adult populations, who face unique barriers including stigma, misinformation, and limited access to youth-friendly services [7].

Within the realm of mHealth, chatbots—computer programs designed to simulate human conversation using text or voice [8]—have emerged as particularly promising tools across a variety of healthcare applications, including diagnosis, treatment, monitoring, and patient rehabilitation for various conditions [9]. Despite this, the use of chatbots for sexual health management, including counseling, and education focusing on African populations remains limited [10]. Further, rule-based chatbots, which operate using a set of predefined rules or decision trees that map specific user inputs to fixed responses, have historically been favored for delivering structured educational content due to their consistency, greater control over responses, and resource-efficient development process [10]. However, these chatbots have faced key limitations, such as difficulty processing dynamic inputs, managing user errors, and operating within narrow knowledge scopes [10]. As a result, their effectiveness in interactive and adaptive scenarios, such as sexual health counseling, where flexibility and nuanced understanding are essential, has been limited.

To address these challenges, we developed STI Shield, a multi-chatbot mobile application that combines rule-based and large language model (LLM)-powered components. The rule-based modules ensure standardized delivery of structured educational content, while the LLM-based module enables adaptive, personalized counseling. Guided by the Health Belief Model (HBM) [11] and Persuasive System Design (PSD) principles [12], the app is designed to support behavior change and reduce risky sexual behaviors in African young adults. Through this novel hybrid design, we aim to offer a more engaging and effective approach to sexual health education compared to traditional chatbot solutions. This paper presents an overview of the design and development of STI Shield, together with initial findings from ongoing user evaluations.

2 Background and Related Work

A recent review of sexual health chatbots [10] found only four chatbot studies focused on improving sexual health among Africans, with just one [13] targeting young individuals. Conducted in Lesotho and focused on women, this study by Nkabane-Nkholongo [13] provided HIV prevention education and reported high acceptability. Participants valued the chatbot's usefulness, privacy, accessibility, lack of time constraints, and nonjudgmental tone.

Recent research has begun to investigate the use of generative chatbots across various healthcare applications, including the dissemination of sexual health information [14]. For instance, a fine-tuned GPT-3.5 chatbot was found to be accurate and safe in providing sexual health information to users, helping reduce healthcare professionals' workload [15] and Snapchat's MyAI offered concise, digestible, and informative responses on topics like consent and sexting [16]. However, these efforts lack structured, theory-driven approaches tailored to high-risk populations, particularly African young adults.

Persuasive systems are software or information systems aimed at influencing attitudes or behaviors without coercion or deception [12]. The PSD principles include four categories: *primary task support*, which helps users achieve goals through methods like *tailoring* and *tunneling*; *dialogue support*, which encourages interaction via feedback mechanisms; *system credibility support*, which builds trust through qualities such as *expertise* and *trustworthiness*; and *social support*, which leverages social influence to shape user behavior [12]. Some PSD principles are especially effective when implemented through interactive methods like gamification and serious games [17]. Gamification involves applying game design elements in non-game contexts to boost user engagement, while serious games are games designed for purposes beyond entertainment [18]. Both gamification and serious games have been widely used to engage users and achieve specific learning or behavior change objectives [19]. Examples of gamification used to apply PSD principles include rewards such as virtual badges, which reinforce dialogue support through immediate positive feedback; and leaderboards and point systems that implement social support by fostering interaction, competition, or collaboration among users [17].

The Health Belief Model (HBM) [11] is a behavioral framework that explains health-related decisions based on individuals' beliefs about a condition and their possible responses. It includes key constructs such as *perceived susceptibility* to a condition, *perceived severity* of its consequences, *perceived benefits* of taking action, and *perceived barriers* to action. According to the HBM, *cues to action* (e.g., reminders) and confidence in one's ability to act (i.e., *self-efficacy*), also play a central role in motivating preventive behavior.

According to Fetrati et al. [10], PSD principles have been widely used in chatbots for sexual education, STI prevention, and treatment adherence, with generally positive outcomes. However, key principles related to games and gamification (e.g., *competition*) remain underutilized. Additionally, few sexual health chatbots are grounded in established behavioral frameworks [10]. For example, the HBM has only been applied in one study on HIV medication adherence among methamphetamine users with mixed results. The outcomes of this study showed reduced methamphetamine use but no improvement in adherence [20]. Given the HBM's strong track record in predicting health behavior change [21], further research is needed to assess its potential in guiding STI-focused chatbot interventions. Similarly, considering prior evidence on the effectiveness of games and gamification in promoting engagement and positive health behaviors among users [22], investigating the integration of underused PSD principles through gamification in STI education contexts and their effectiveness is warranted.

3 Design and Development of STI Shield

STI Shield was developed as a native Android application comprising five core modules (Fig. 1):

1. *Knowledge Quest*: A rule-based chatbot featuring a text-based game set in an ancient African storyline (Fig. 1a). Users act as a warrior recovering lost Scrolls of Health containing information about STIs, including their symptoms, transmission, treatments, and prevention.
2. *STI Smart*: A rule-based chatbot presented by African volunteer personas hosting quiz games to test users' STI knowledge (Fig. 1b).
3. *Health Guard*: A rule-based chatbot offering risk assessment and a personalized STI testing and vaccination plan based on user personal sexual history, presented by a customizable African doctor persona.
4. *Safe Talk*: An LLM-based generative chatbot with a customizable African doctor persona (Fig. 1c). The chatbot provides counseling on topics such as STIs, sexual dysfunction, orientation, contraception, consent, and access to care facilities (e.g., HIV clinics).
5. *Leaderboard and Point System*: Users earn points by completing in-app activities, and their scores are displayed on a competitive leaderboard that ranks their performance relative to others.

The app's, rule-based chatbots were custom-built using decision tree structures. The generative chatbot in the Safe Talk module was developed using OpenAI's Assistants API [23]. Educational and advisory content was drawn from reputable sources, including the Public Health Agency of Canada [24], the Centers for Disease Control and Prevention [25], and the World Health Organization [26]. Quiz questions for the STI Smart module were primarily adapted from the STD Knowledge Questionnaire [27] and the HIV Knowledge Questionnaire [28], with additional items based on information from the aforementioned public health sources.

Ethical and safety concerns were mitigated through several measures: User conversations with the chatbots were neither stored nor reused for training purposes, and they were not accessible to third parties. All conversation data were deleted from both the device and cloud once each chat session ended. Sources were cited within responses so users could verify the legitimacy of information. In sensitive cases (e.g., abuse, trauma), the chatbot directed users to seek professional help. These safeguards were explained in the app's privacy policy and reiterated in conversations where relevant. In addition, the app allowed users to directly contact the research team with any concerns regarding privacy, data handling, chatbot responses, or other issues.

Further, each module was implemented at the intersection of various HBM and PSD principles. For instance, the HBM construct of *perceived susceptibility* was addressed by highlighting the risks associated with common sexual practices that users may view as normal or harmless, such as unprotected oral sex, and by presenting STI prevalence statistics specific to the user's age group to contextualize those risks. The PSD principle of *tailoring* was implemented throughout the app by incorporating African-themed elements, including culturally relevant chatbot personas, game storylines, and imagery. The app also incorporated other PSD principles. For example, the *rewards* principle was applied by awarding in-app points for completing actions such as finishing each level in the Knowledge Quest or completing an assessment in the Health Guard module, while the *competition* principle was implemented through the app's leaderboard feature.

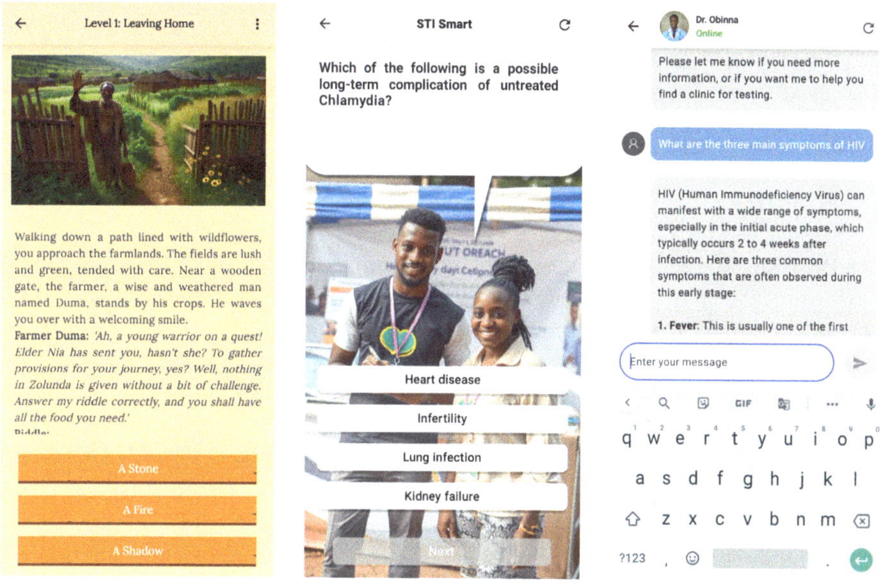

(a) Knowledge Quest (b) STI Smart (c) Safe Talk

Fig. 1. Screenshots of application modules.

4 Evaluation

Following pilot evaluations of STI Shield [29] and iterative enhancements based on user feedback, a large-scale mixed-methods study is underway. Recruitment commenced after receiving approval from Dalhousie University's Health Sciences Research Ethics Board. All participants provided informed consent prior to participation, completed a pre-test questionnaire, used the app for seven days, and then completed a post-test questionnaire, followed by an optional interview to provide in-depth feedback on their experience. The aims of the study include evaluating (1) the perceived persuasiveness of the app in motivating safer sexual practices and reducing risky behaviors; (2) its impact on users' STI knowledge; and (3) user perceptions of the usability and acceptability of the app.

Perceived persuasiveness was assessed using the Perceived Persuasiveness Scale developed by Orji et al. [30], which uses a 5-point Likert scale. A one-sample t-test was conducted to determine whether scores significantly differed from the scale midpoint (3), and an independent samples t-test examined differences based on participant genders. Changes in STI knowledge were assessed by comparing pre- and post-test quiz scores on the STI Smart module using paired-samples t-tests across three quiz difficulty levels: easy, medium, and hard. Quiz scores were calculated out of 100. Usability was evaluated using the System Usability Scale (SUS) [31], and interpreted using benchmarks from Bangor et al. [32] and Sauro et al. [33]. Bangor et al. [32] classify SUS scores from *Worst Imaginable* and *Poor*, through the midpoint of *OK*, to more positive ratings of *Good*, *Excellent*, and *Best Imaginable*. Sauro et al. [33] complement this by mapping scores onto an $A+$ to F grading scale. Additionally, an independent samples t-test was conducted to examine differences in SUS scores across participant genders.

5 Results

Results from 40 African young adults (20 males and 20 females, aged 18–30 years) indicate that participants rated the app as highly persuasive (M = 4.63, SD = 0.33), significantly above the neutral midpoint, $t(39)$ = 31.73, p <.001. No significant gender differences were found in persuasiveness ratings, with females reporting M = 4.60 (SD = 0.35) and males M = 4.66 (SD = 0.31), $t(38)$ = -0.60, p = .550.

Moreover, participants demonstrated significant gains in STI knowledge across all three quiz difficulty levels. For easy questions, scores increased from M = 76.50 (SD = 27.32) to M = 99.00 (SD = 4.41), $t(39)$ = 5.37, p <.001. For medium questions, scores rose from M = 93.50 (SD = 9.21) to M = 99.25 (SD = 3.50), $t(39)$ = 4.03, p <.001. For hard questions, scores improved from M = 79.00 (SD = 23.40) to M = 97.75 (SD = 11.21), $t(39)$ = 5.45, p <.001.

Lastly, the app was rated highly usable, with an average SUS score of 90.13 (SD = 5.43) out of 100. According to the benchmark developed by Bangor et al. [32], this score falls within the *Excellent* adjective rating category and is positioned at the high end of that range. Additionally, based on the curved grading scale proposed by Sauro et al. [33], this score corresponds to an $A+$, which is the highest possible grade. Moreover, female participants reported slightly higher usability (M = 91.75, SD = 4.52) than males (M = 88.50, SD = 5.87); however, this difference was not statistically significant, $t(38)$ = 1.96, p = .057.

6 Discussion and Conclusion

Our findings provide promising early evidence that STI Shield effectively engages African young adults with persuasive and user-friendly sexual health education. The knowledge gains across all quiz difficulty levels highlight the app's ability to convey both basic and advanced STI information. In addition, high perceived persuasiveness ratings demonstrate the potential of STI Shield in successfully motivating users to adopt safer sexual behaviors. Furthermore, the excellent usability rating indicates strong potential for widespread acceptance and sustained use. Overall, these results underscore the effectiveness of our hybrid approach, which integrates rule-based and LLM-driven modules grounded in the HBM and PSD principles. Gamification enhances the delivery of these principles by fostering increased engagement, and sustaining user motivation. Our ongoing analyses will build on these initial findings, aiming to better understand how digital tools can enhance sexual health education and encourage positive behavior change within healthcare contexts in underserved populations.

References

1. Kassie, B.A., Yenus, H., Berhe, R., Kassahun, E.A.: Prevalence of sexually transmitted infections and associated factors among the university of gondar students, northwest ethiopia: a cross-sectional study. Reprod. Health **16**, 1–8 (2019)
2. UNESCO. Comprehensive sexuality education: For healthy, informed and empowered learners (2025). https://www.unesco.org/en/health-education/cse. Accessed 10 Jan 2025
3. GSMA Intelligence. The mobile economy Sub-Saharan Africa 2020 (2020). https://www.gsma.com/solutions-and-impact/connectivity-for-good/mobile-economy/wp-content/uploads/2020/09/GSMA_MobileEconomy2020_SSA_Eng.pdf. Accessed 02 Mar 2025
4. Tarricone, R., Petracca, F., Ciani, O., Cucciniello, M.: Distinguishing features in the assessment of mhealth apps. Expert Rev. Pharmacoecon. Outcomes Res. **21**(4), 521–526 (2021)
5. Onukwugha, F.I., et al.: The effectiveness and characteristics of mhealth interventions to increase adolescent's use of sexual and reproductive health services in sub-saharan africa: a systematic review. PLoS ONE **17**(1), e0261973 (2022)
6. Jones, C., Cui, Y., Jeminiwa, R., Bajracharya, E., Chang, K., Ma, T.: Personalized and culturally tailored features of mobile apps for gestational diabetes mellitus and their impact on patient self-management: Scoping review. JMIR Diab. **9**, e58327 (2024)
7. Harris, B., et al.: Inequities in access to health care in South Africa. J. Public Health Policy **32**, S102–S123 (2011)
8. Adamopoulou, E., Moussiades, L.: Chatbots: history, technology, and applications. Mach. Learn. Appl. **2**, 100006 (2020)
9. Laranjo, L., et al.: Conversational agents in healthcare: a systematic review. J. Am. Med. Inf. Assoc. **25**(9), 1248–1258 (2018)
10. Fetrati, H., Chan, G., Orji, R.: Chatbots for sexual health improvement: a systematic review. Int. J. Human-Comput. Interact. 1–23 (2024)
11. Champion, V.L., Skinner, C.S., et al.: The health belief model. Health Behav. Health Educ. Theory Res. Pract. **4**, 45–65 (2008)
12. Oinas-Kukkonen, H., Harjumaa, M.: Persuasive systems design: key issues, process model, and system features. Commun. Assoc. Inf. Syst. **24**(1), 28 (2009)
13. Nkabane-Nkholongo, E., Mokgatle, M., Bickmore, T., Julce, C., Jack, B.W.: Adaptation of the gabby conversational agent system to improve the sexual and reproductive health of young women in lesotho. Front. Digital Health **5**, 1224429 (2023)

14. Moulaei, K., Yadegari, A., Baharestani, M., Farzanbakhsh, S., Sabet, B., Afrash, M.R.: Generative artificial intelligence in healthcare: a scoping review on benefits, challenges and applications. Int. J. Med. Inf. 105474 (2024)
15. Latt, P.M., et al.: Real-world evaluation of artificial intelligence (ai) chatbots for providing sexual health information: a consensus study using clinical queries (2024)
16. Marcantonio, T.L., Avery, G., Thrash, A., Leone, R.M.: Large language models in an app: conducting a qualitative synthetic data analysis of how snapchat's "my ai" responds to questions about sexual consent, sexual refusals, sexual assault, and sexting. J. Sex Res. 1–15 (2024)
17. Krath, J., Von Korflesch, H.F.: Designing gamification and persuasive systems: a systematic literature review. In: GamiFIN, pp. 100–109 (2021)
18. Deterding, S., Dixon, D., Khaled, R., Nacke, L.: From game design elements to gamefulness: defining "gamification". In Proceedings of the 15th International Academic MindTrek Conference: Envisioning Future Media Environments, pp. 9–15 (2011)
19. Sailer, M., Homner, L.: The gamification of learning: a meta-analysis. Educ. Psychol. Rev. **32**(1), 77–112 (2020)
20. Moore, D.J., et al.: Individualized texting for adherence building (ITAB) for methamphetamine users living with HIV: a pilot randomized clinical trial. Drug Alcohol Depend. **189**, 154–160 (2018)
21. Jones, C.J., Smith, H., Llewellyn, C.: Evaluating the effectiveness of health belief model interventions in improving adherence: a systematic review. Health Psychol. Rev. **8**(3), 253–269 (2014)
22. Ndulue, C., Orji, R.: Games for change—a comparative systematic review of persuasive strategies in games for behavior change. IEEE Trans. Games **15**(2), 121–133 (2022)
23. OpenAI. Assistants API overview (2025). https://platform.openai.com/docs/assistants/overview. Accessed 10 Jan 2025
24. Public Health Agency of Canada. Booklet: Sexually transmitted infections (2025). https://www.canada.ca/en/public-health/services/publications/diseases-conditions/booklet-sexually-transmitted-infections.html. Accessed 10 Jan 2025
25. Centers for Disease Control and Prevention. Sexually transmitted diseases treatment guidelines: Primary prevention methods (2024). https://www.cdc.gov/std/treatment-guidelines/clinical-primary.htm. Accessed 10 Jan 2025
26. World Health Organization. Global sexually transmitted infections programme (2025). https://www.who.int/teams/global-hiv-hepatitis-and-stis-programmes/stis/prevention. Accessed 10 Jan 2025
27. Jaworski, B.C., Carey, M.P.: Development and psychometric evaluation of a self-administered questionnaire to measure knowledge of sexually transmitted diseases. AIDS Behav. **11**, 557–574 (2007)
28. Carey, M.P., Morrison-Beedy, D., Johnson, B.T.: The HIV-knowledge questionnaire: development and evaluation of a reliable, valid, and practical self-administered questionnaire. AIDS Behav. **1**, 61–74 (1997)
29. Fetrati, H., Chan, G., Orji, R.: Leveraging generative and rule-based models for persuasive STI education: a multi-chatbot mobile application. In: Proceedings of the 7th ACM Conference on Conversational User Interfaces, pp. 1–9 (2025)
30. Orji, R., Reilly, D., Oyibo, K., Orji, F.A.: Deconstructing persuasiveness of strategies in behaviour change systems using the arcs model of motivation. Behav. Inf. Technol. **38**(4), 319–335 (2019)

31. Brooke, J., et al.: Sus-a quick and dirty usability scale. Usabil. Eval. Ind. **189**(194), 4–7 (1996)
32. Bangor, A., Kortum, P., Miller, J.: Determining what individual sus scores mean: adding an adjective rating scale. J. Usabil. Stud. **4**(3), 114–123 (2009)
33. Sauro, J., Lewis, J.R.: Quantifying the user experience: Practical statistics for user research. Morgan Kaufmann (2016)

AI Chatbots in Higher Education: Students' and Teachers' Perceptions of Challenges and Good Practices for Teaching

Konstantinidou Sofia, Ioannis Lefkos(✉), and Nikolaos Fachantidis

Laboratory of Informatics and Robotics in Education and Society, University of Macedonia, Thessaloniki, Greece
lefkos@uom.edu.gr

Abstract. The study examines the perceptions of higher education students and teachers regarding challenges and provides good teaching practices by teachers for an effective integration of AI technology in education. The research utilises qualitative data from an open question to 372 students and unstructured interviews with 10 teachers. The findings highlighted challenges by using chatbots in education, such as low reliability of information, impact on the job industry, and on essential learning from both, with teachers adding more to them, including possible solutions. To enhance the effective integration of these technologies in higher education, teachers emphasised that the best teaching practice is to encourage students to use them while critically evaluating the information based on their knowledge and other learning materials. These insights can be valuable in guiding policymakers and other stakeholders in overcoming obstacles and fostering pedagogical practices.

Keywords: AI in Education · Chatbots · Higher Education · Students · Teachers

1 Literature Review

Artificial Intelligence (AI) is a technology related to computer systems that imitates human behaviour and intelligence [12]. Generative AI, in particular, not only mimics human intelligence but also generates original content [7]. In this subfield, chatbots or conversational agents are included, that seem to have attracted considerable attention, taking into consideration the emergence of ChatGPT [16], which led the way for continuously evolving chatbots and many related applications being developed.

Adding to the above, it can be supported that generative artificial intelligence has had a significant impact on many factors of everyday life, as well as on teaching and learning, from preschool to higher education. Specifically, there seems to be a strong interest in using chatbots in educational settings [20] as they provide learning assistance to both teachers and students [13].

Given the fact that it is an easy and accessible technology [3], chatbots can facilitate learning and adopt various roles in higher education contexts. They can assist students

M. E. Auer et al. (Eds.): IMCL 2025, LNNS 1954, pp. 176–185, 2026.
https://doi.org/10.1007/978-3-032-24998-2_15

in administrative [15] and learning tasks [1], increase their self-regulation and interest for learning [21].

From a global perspective, chatbots' learning outcomes have been adequately demonstrated from relevant research. However, there are regions where the integration of AI is progressing more slowly, and the field of teachers' perceptions regarding these technologies remains underexplored. An example of this state is Greece, where AI in education is still emerging. Hence, this provides an opportunity to explore the perceptions of future teachers and teachers in a region still undergoing digital transformation, while at the same time contributing to the global discussions on the implementation of AI education.

Turning to the Greek context, there is a limited number of studies concerning the deployment of chatbots in education, with the majority concentrating on students' perceptions within primary or secondary educational settings [2, 11].

Research investigating the perceptions of educators and learners regarding the use of chatbots in educational environments has demonstrated an overall positive view and inclination to utilise these technologies [4, 5]. However, as with any other technology, the coin has two sides. The responders express concerns about the potential future impact of these tools [5] and the need for adequate user training [20]. Furthermore, the potential for over-reliance on these tools and the consequent reduction in creativity and critical thinking within the educational process make them feel even more concerned, as educators [4].

To the best of our knowledge, regarding the Greek context, only a few studies are oriented towards higher education settings. For example, [10] conducted a survey involving 515 students and gathered their views about the utilization of AI tools. The results showed, on the one hand, students' acknowledgement of their potential in research and in affording personalised learning experiences, but on the other hand, their concerns regarding the ethical use, the trustworthiness of the results and the likelihood of depreciation of critical thinking skills. Exhibiting quite similar findings, [19] surveyed 231 future educators and explored their knowledge, their willingness to use and their concerns, while at the same time compared undergraduates' and master's level students' perceptions. They found that students involved in hands-on experiences with AI tools expressed fewer concerns and more equalized perceptions.

This research analyses university students' and teachers' perceptions on challenges about chatbots' usage in higher education and provides good practices given by teachers. The study identifies key challenges and teaching methods that could address the challenges and enhance an effective integration of AI in higher education. Higher education attracts considerable research interest due to the necessity of exploring innovative and effective teaching methodologies, considering the increased specialization required.

Contributing to the significance of our study, our research shows and compares students' and teachers' perceptions of challenges by using chatbots in education while presenting recommendations and good practices for an effective integration in educational contexts from experts in education.

2 Research Methodology

This study explores the integration of chatbots within a higher education setting by analysing data collected from higher education students/future teachers and higher education teachers.

The study is driven by the following research questions (RQs):

RQ1: What are students' and teachers' perceptions regarding the challenges of using AI chatbots such as ChatGPT?

RQ2: Are there any differences between students' and teachers' perceptions regarding the challenges of using AI chatbots such as ChatGPT?

RQ3: What practices do teachers recommend for the effective integration of chatbots in higher education?

A non-probability, convenience sampling method was used involving voluntary participation, to effectively collect data within the available time and resources. The sample comprises, on the one hand, 372 students / future teachers registered in a Greek university pedagogical course, and on the other hand, 10 higher education teachers working in Greek universities.

The research tool used to collect data from students was a structured questionnaire adapted from [4], which assessed the following: (a) knowledge; (b) willingness to use; (c) concerns; (d) the perceived benefits and challenges of AI technologies such as ChatGPT with both closed and open-ended questions. In this research, we present the results that emerged from the open-ended questions, while our findings from the closed-ended questions were reported elsewhere [9]. Therefore, the questionnaire consists of four (4) demographic questions and two (2) open-ended questions that provide a deeper insight into the respondents' perceptions.

The structure of the questionnaire was essential for providing the data required for our study. The tool was used in its original form, along with three (3) additional demographic questions regarding the participants' gender, study semester, and the frequency of AI usage.

The two (2) open questions included in the questionnaire aimed to explore the reasons behind students' willingness to use chatbots such as ChatGPT, as well as the reasons for having concerns or a lack thereof about AI technologies. For the purposes of our study, in this paper, we utilised one (1) of the open-ended questions that collected participants' concerns about using chatbots such as ChatGPT.

Data was collected digitally via a questionnaire, which was securely stored on an online hard drive accessible solely by the researchers until the conclusion of the study. All participants provided their informed consent and agreed to participate after being thoroughly briefed on the study's objectives, procedures, and potential risks benefits. Furthermore, this study was approved by the university's Research Ethics Committee.

Regarding the perceptions of teachers, we employed a qualitative approach as well, using unstructured interviews with university teachers. The interviews were conducted in person or online, based on teachers' availability.

After collecting the data, a thematic analysis of the qualitative data was conducted. All responses were reviewed and examined to fully understand and categorize emerging themes. In the first stage, the first author identified emerging categories, drawing on themes from participants' answers and existing research on similar topics, and provided detailed descriptions and representative quotes. The categorisation scheme was developed based on themes from the original article [4] after making relevant modifications.

Subsequently, the two additional authors independently classified 30% of the responses. All three authors addressed the discrepancies in the discussion section and reached a consensus, ensuring no unresolved issues remain.

In the following reporting of results, we chose to include only categories with more than ten (10) responses, as these signify the most frequently mentioned themes.

3 Results

Table 1 shows that the research sample of students (N = 372) mostly comprises women (N = 326), in their 1st semester of studies (N = 264). Regarding the frequency of AI usage among participants, high variability is observed, with a significant percentage of never/rarely using AI technologies (N = 224), and another percentage of students using these technologies sometimes/often/always (N = 148).

Table 1. Demographic characteristics of higher education students

	N	Percentage (%)
Gender		
Men	46	12.4
Women	326	87.6
Semester		
1st semester	264	71.0
2nd semester and above	108	29.0
Frequency of AI usage		
Low usage [never, rarely]	224	60.2
High usage [sometimes, often, always]	148	39.8

Table 2 reveals that the research sample of higher education teachers predominantly consists of men that combine their specialty with technology science and have high familiarity and AI usage in their daily and educational practices.

Table 3 shows the thematic analysis which revealed reasons for students' concerns about chatbots such as ChatGPT, along with perceived challenges in descending order.

The primary concern among students is the accuracy and reliability of information provided by generative AI technologies. Most of them worry that responses given by the chatbot may not always be objective and could contain biased or misleading information. Many talked about "*insufficient information*", "[…] *inaccurate or wrong information*". There is also scepticism about the possibility of outdated and out of-context responses, which makes students question their trust in AI content, by saying that "*they cannot know how timely they are and whether they are in line with society's frameworks*".

Table 2. Demographic characteristics of higher education teachers

Participant	Gender	Specialty	Familiarity and AI usage
P1	Woman	English Language, Methodology of Teaching and Technology	High
P2	Woman	Psychology and Human-Robot Interaction	High
P3	Man	Computer Networks	Moderate
P4	Man	Computer Networks	High
P5	Woman	Science Teaching and Educational Technology	Moderate
P6	Man	Economics and Informatics	Low
P7	Man	Internet Politics	Low
P8	Man	Physical Sciences and Computer Science in Education	High
P9	Man	Science Teaching and Educational Technology	Low
P10	Man	Science Teaching and AI in Education	High

Another concern they had was the potential impact on students' skills development such as critical thinking, creativity, problem-solving and cooperation. As one noted for long-term consequences of this, "*future generations won't use their critical thinking to solve problems and they won't learn to cooperate*", while another added that "*the dependence of research and completion of tasks through artificial intelligence and the alienation of group tasks of communication, emotions expression and socialization.*".

Also, the convenience AI provides can reduce the quality of the learning process because students won't engage with the material. They often bypass critical steps of analysing, synthesising, and processing information, leading to shallow learning. As one noted "*students copy the answers*", while other said that "[…] *information is not always filtered in the right way, and this excessive convenience can lead to a lack of methodical search for information*". Overall, this results in "[…] *the absence of authenticity, meaningful effort*".

Table 4 shows the thematic analysis for the challenges that result from using chatbots in higher education as perceived by teachers.

Participants expressed challenges related to the low reliability of answers given by chatbots. Some of them focused on the fact that some of them do not provide their source of information when giving an answer, as P3 said "*We take a short text like a story, put together in a way that makes grammatical sense, but you're not sure what sources it came from*". The participants then emphasized on solutions like P8 who said that "*we need to learn more about the tool, how it works and what its basic limits are*" and P1 who noted that "*we need to be more specialized in the part where we use it than it is*".

Table 3. Students' reasons of concerns about chatbots such as ChatGPT

Themes	Subthemes
Accuracy and reliability of information	● Concerns about AI providing answers that are not objective or might be biased ● Worries about whether the information is valid, truthful, and up to date ● Scepticism regarding the possibility of incorrect or outdated responses
Impact on skills development	● AI may diminish students' ability to put in effort, affecting their development of essential skills ● Reduced engagement in problem-solving, thinking critically, and using creativity ● Reduced opportunities for teamwork and collaborative learning
Impact on essential learning	● The ease of access to information reduces the quality of the learning process ● Relying on AI for assignments will not encourage students to process information deeply or engage in meaningful learning ● AI-driven assignments might lack originality and personal thought, making learning less authentic

Many of them also believed that there could be a potential impact in the job industry with many jobs being replaced by AI. Although they supported that, they included in their answers that many other jobs could be created. As P8 noted "*It is certain that some professions may no longer be necessary, while others may be created*" with P5 agreeing that "*society adapts to technological advancements, it's a two-way relationship, so I think that maybe some jobs will indeed be reduced, but I think some jobs will also be created*".

Ethical issues were also raised about the way students use these technologies which affects their learning and academic integrity. P4 contributes these to the fact that "*many students use these tools very superficially, very shallowly, without checking whether what these tools give them is valid*". Many of them also supported that these issues existed before this technology, even with traditional tools. P8 said that "*the difference now is perhaps that instead of asking for help from a classmate or someone else, anyone can very easily ask for help from their mobile phone. So people's behaviour hasn't essentially changed, it's the same tool that has changed*".

Participants find that sometimes students use chatbots in a wrong way which is caused by their lack of awareness. As P6 said, "*you get an answer but you're not in a position to evaluate whether it's good or not. I think that's where the insecurity comes from*". Education on these technologies should be given not only to students but teachers as well, as P2 notes "*Teachers must also be trained to educate students and students must also learn how to use and integrate it*".

Some of them expressed that skills like writing could be reduced from the usage of AI. Although, they supported that new skills are developed too. As P10 noted, "*of course skills are being reduces. But that doesn't mean that we can't focus on others and move forward*". P8 said that "*technology forces us to change skills and develop some other skills. And maybe those skills are of higher order than what we needed for the previous generation of technology*".

One participant expressed the challenge of dependence on the technology based on the fact that many people cannot write without it. As P1 said,"*I mean, I need it so much when a year ago it didn't exist in my life*".

Table 4. Teachers' perceived challenges by AI usage in education

Themes	Responses
Low reliability of answers and information	P1, P2, P3, P4, P5, P6, P7, P9
Impact in the job industry	P1, P2, P4, P5, P6, P8, P9, P10
Ethical issues	P1, P2, P3, P4, P6, P7, P8, P9
Wrong use due to lack of awareness	P1, P2, P3, P5, P6, P7, P9
Impact on skills development	P1, P10
Overuse and dependence	P1

Table 5 shows the thematic analysis for the benefits that result from using chatbots in higher education as perceived by teachers.

Participants supported that AI usage will contribute to new skills and critical thinking development. As they said, chatbots are technology that continues to evolve and therefore, so the people that use them by checking the information. As P8 said "*technology is making us change skills and develop other skills. And maybe these skills are of higher level as compared to those we needed for the previous technology generation*".

Others supported that chatbots contribute to providing fast answers for a wide range of topics like a search engine. As P4 says "*I think it's an excellent technology for a student to have fast answers in a wide range of subjects but not answers that need technical knowledge*". In fact, they provide them with innovative ideas that they have not thought before and can adopt accordingly. As P2 said "*it pointed out a new parameter that I have not accounted for.* [...] *So, then I searched about it and later I incorporated this factor in my report*". Lastly, small number of participants supported the facilitation in administrative issues, with P1 suggesting that "*the chatbot answered to an e-mail which is bureaucratic and answers really well in an appropriate style*".

Table 6 shows the recommendations and good practices given by teachers to have an effective integration of chatbots in higher education.

Table 5. Teachers' perceived benefits by AI usage in education

Themes	Responses
1. Development of new skills and critical thinking	2. P5, P7, P8, P9, P10
3. Fast answers in a wide range of topics	4. P4, P3, P7, P8
5. Innovative ideas	6. P2, P6, P8
7. Assist in administrative issues	8. P1, P2

Table 6. Teachers' recommendations and good practices of AI usage in higher education

Recommendations	Responses
Teachers can encourage students to use chatbots to facilitate their learning tasks and research but be able to explain the solution given, its limitations, its strengths, and how it could be changed using other learning sources too	P4, P5, P6, P7, P8, P9
Students can use chatbots to be more creative and get ideas on a subject while still using the provided educational material given by the teacher	P1
Teachers, as role models, can use chatbots during class to teach students how they work and how they can interact with them	P2
Specialized people in AI can create personalized chatbots for university classes which can be used by students to answer questions related to the subjects	P3
Teachers can encourage students to use any chatbot they want and create an educational scenario where the chatbot could be used but be able to acknowledge its limitations, strengths and the added educational value	P5
Teachers can utilise learning methods such as flipped classroom, self-regulated learning and personalized feedback by encouraging students to have a chat with a chatbot to learn something new before their actual lesson	P10

4 Discussion and Conclusions

The study offers insights into the impact of AI chatbots in higher education, focusing on the challenges perceived by students and teachers and good practices that could lead to effective learning using chatbots. Students supported that chatbots' answers could be inaccurate and unreliable, jobs could be replaced by automation and learning loses its essential purpose. Similar challenges are being supported by teachers too such as low reliability of answers and information, impact in the job industry, but also adding challenges related to ethical issues, wrong usage due to lack of awareness, impact on skills development and potential overuse and dependence. These challenges align with the research by [4, 8, 14, 17, 18].

Looking at the answers given by students and teachers, similarities and differences can be found. First, both focused on the fact that these systems make mistakes and need to be evaluated, and they could affect the job market by replacing some jobs. Moving on to the differences, teachers seem to have a more comprehensive opinion on technology and its limitations, adding more challenges, while also being able to explain how each of them could be addressed. While answering the challenges, they also provided possible solutions focusing on increasing the pedagogical benefits of technological advancement.

Teachers expressed benefits by using chatbots in higher education as they support new skills development, provide fast answers in a wide range of knowledge, innovative ideas and assist in administrative issues. These benefits align with other research, such as [4, 8, 14].

Teachers were also asked to give recommendations on good practices that could enhance the effectiveness of chatbots usage in higher education learning. The most effective one for most of them was the encouragement of using chatbots in academic tasks and being able to critically approach the answers given by researching and studying other learning materials too. This exploratory approach seems to be effective, as they develop critical thinking skills to assess the information provided while being allowed to use it in their academic obligations. This practice has been proven to be particularly effective in equipping learners with critical thinking skills during their interaction with artificial intelligence tools [6].

Although this study is centered on Greece, its findings significantly contribute to the wider international discourse on artificial intelligence in education. Gaining insights into the perceptions of students and teachers across diverse contexts assists in shaping global strategies for AI-enhanced learning. The perceived challenges and recommendations for teaching can serve as valuable guidance for policymakers and other stakeholders in the development of ethical guidelines and programs. These initiatives should aim to overcome obstacles and promote pedagogical practices that are effectively integrated.

Future research could also explore how students perceive teachers who incorporate AI technologies, including chatbots, into their pedagogical practices, in comparison to those who rely solely on traditional educational technologies.

Acknowledgements. This study was funded by the project "SOCIOBOT4SYNERGY: Optimising Education and Well-Being through Collaborative Social Robotics," coordinated by the Research Committee of the University of Macedonia.

References

1. Alsadoon, R.: Chatting with AI bot: vocabulary learning assistant for Saudi EFL learners. Engl. Lang. Teach. **14**(6), 135–157 (2021)
2. Athanassopoulos, S., Manoli, P., Gouvi, M., Lavidas, K., Komis, V.: The use of ChatGPT as a learning tool to improve foreign language writing in a multilingual and multicultural classroom. Adv. Mobile Learn. Educ. Res. **3**(2), 818–824 (2023)
3. Castillo, A.G., et al.: Effect of Chat GPT on the digitized learning process of university students. J. Namibian Stud. History Politics Culture **33**, 1–15 (2023)
4. Chan, C., K., Y., Hu, W.: Students' voices on generative AI: perceptions, benefits, and challenges in higher education. Int. J. Educ. Technol. Higher Educ. **20**(43) (2023)

5. Ghotbi, N., Ho, M., T., Mantello, P.: Attitude of college students towards ethical issues of artificial intelligence in an international university in Japan. AI & SOCIETY **37**, 283–290 (2022)
6. Guo, K., Zhong, Y., Li, D., Chu S., K., W., C.: Investigating students' engagement in chatbot supported classroom debates. Interactive Learning Environ. **31**(5) (2023)
7. Karpathy, A., et al.: Generative models. OpenAI (2016). https://openai.com/index/generative-models/
8. Kasneci, E., et al.: ChatGPT for good? On opportunities and challenges of large language models for education. Learning and Individual Differences **103** (2023)
9. Konstantinidou, S., Lefkos, I., Fachantidis, N.: Are they AI-competent? Future Teachers' Readiness to Use Conversational Agents as Learning Assistants. Int. J. Info. Educ. Technol. **16**(2), 318–329 (2026)
10. Kostas, A., Paraschou, V., Spanos, D., Tzortzoglou, F., Sofos, A.: AI and ChatGPT in Higher Education: Greek Students' Perceived Practices, Benefits, and Challenges. Education Sciences **15**(5), 605 (2025)
11. Kotsis, K.T.: ChatGPT in teaching physics hands-on experiments in primary school. European J. Educ. Stud. **11**(10), 123–143 (2024)
12. McCarthy, J.: What is Artificial Intelligence (2007)
13. Mendoza, S., et al.: A model to develop chatbots for assisting the teaching and learning process. Sensors **22**(15) (2022)
14. Ngo, T., T., A.: The perception by university students of the use of ChatGPT in education. Int. J. Emerg. Technol. Learn. **18**(17), 4–19 (2023)
15. Nguyen, T.H., Waizenegger, L., Techatassanasoontorn, A.: Don't neglect the user – identifying types of human-chatbot interactions and their associated characteristics. Info. Sys. Front. **24**, 797–838 (2021)
16. OpenAI (2022). Introducing ChatGPT. OpenAI. https://openai.com/index/chatgpt/
17. Perez, J.S., Daradoumis, T., Puig, J.M.M.: Rediscovering the use of chatbots in education: a systematic literature review. Comp. Appl. Eng. Educ. **28**(6), 1549–1565 (2020)
18. Rane, N.L., Choudhary, S.P., Tawde, A., Rane, J.: ChatGPT is not capable of serving as an author: ethical concerns and challenges of large language models in education. Int. Res. J. Modernization in Eng. Technol. Sci. **5**(10), 851–874 (2023)
19. Tsiani, M., Lefkos, I., Fachantidis, N.: Perceptions of Generative AI in Education: Insights from Undergraduate and Master's-Level Future Teachers. Int. J. Pedagogical Res. **9**(2), 89–108 (2025)
20. Tsivitanidou, O., Ioannou, A.: Envisioned Pedagogical Uses of Chatbots in Higher Education and Perceived Benefits and Challenges. In: Zaphiris, P., Ioannou, A. (eds.) Proceedings of the Conference on Learning and Collaboration Technologies: Games and Virtual Environments for Learning, pp. 230–250. Springer, Cham (2021)
21. Yin, J., Goh, T., Yang, B., Xiaobin, Y.: Conversation technology with micro-learning: the impact of chatbot-based learning on students learning motivation and performance. J. Educ. Comp. Res. **59**(1), 154–177 (2021)

Learning Analytics

Extending Clustering and Causal Analysis Techniques for Identifying and Explaining Learning Gaps in MOOCs

Rekha Agali Virupaksha(✉), Stefania Zourlidou, and Frank Hopfgartner

University of Koblenz, 56070 Koblenz, Germany
rekha_av@uni-koblenz.de
https://www.uni-koblenz.de/de/informatik/west

Abstract. Massive Open Online Courses (MOOCs) offer broad access to learning but face persistent challenges related to learner disengagement and uneven performance. This study investigates how clustering and anomaly detection techniques can be adapted to identify patterns indicative of learning gaps in MOOCs. Using engagement data from an open dataset, we apply hybrid-weighted K-Means, OPTICS, and Local Outlier Factor (LOF) to detect distinct learner profiles and behavioral anomalies. The results reveal that specific clusters exhibit extremely high proportions of underperforming learners (up to 100%), while LOF identifies outliers with erratic engagement patterns tied to non-completion. Temporal analyses further show steep declines in participation over time. These findings demonstrate the potential of combining clustering and temporal modeling to uncover actionable insights into learning gaps, supporting targeted interventions in large-scale online education.

1 Introduction

Massive Open Online Courses (MOOCs) have revolutionized the delivery of education by offering scalable, flexible, and open-access learning opportunities to a global audience. They have significantly broadened participation in higher education, enabling diverse groups of learners—from working professionals to students supplementing formal curricula—to acquire new skills and knowledge [2,13]. Despite these benefits, MOOCs still face important challenges, especially when it comes to learning gaps.

Learning gaps in MOOCs refer to discrepancies between the intended educational objectives of a course and the actual learning outcomes achieved by participants. These gaps may emerge during the course as learners struggle with complex concepts or engage inconsistently with course material, or persist after the course ends as learners fail to retain or apply acquired knowledge [2,3]. Contributing factors include the heterogeneity of learners' backgrounds, the asynchronous and self-paced nature of MOOCs, and often limited opportunities for personalized instructional support [3,4]. Analyzing and addressing

M. E. Auer et al. (Eds.): IMCL 2025, LNNS 1954, pp. 189–200, 2026.
https://doi.org/10.1007/978-3-032-24998-2_16

these gaps is further complicated by the nature of MOOC data. MOOCs generate large-scale datasets characterized by high dimensionality, encompassing a rich array of engagement metrics such as clickstreams, video interactions, quiz attempts, and forum posts [5]. These datasets also tend to be noisy and sparse due to sporadic participation and high dropout rates, with course completion rates typically averaging around 12% [2]. Such characteristics pose challenges for standard clustering and analytical techniques that aim to uncover meaningful patterns indicative of learning gaps.

Recent work in Educational Data Mining (EDM) has highlighted the promise of clustering algorithms, such as K-Means and DBSCAN, in grouping learners based on engagement patterns and identifying at-risk populations [1,3,16]. However, these methods are often applied in generic forms that do not fully account for the unique dynamics of MOOC environments—including asynchronous participation, temporal shifts in engagement, and diverse learner trajectories [9,10]. Moreover, there remains a critical need to advance beyond purely descriptive clustering toward approaches that can identify specific engagement patterns linked to learning gaps, thereby enabling data-driven interventions.

Against this backdrop, this paper investigates how clustering algorithms can be tailored and extended to effectively identify patterns of learning gaps in MOOCs. Specifically, we address the following research questions:

1. *RQ1:* How can clustering algorithms, such as K-Means and DBSCAN, be optimized to identify patterns of learning gaps in engagement data from MOOCs?
2. *RQ2:* What specific patterns in learner engagement within MOOCs indicate learning gaps, and how can clustering techniques validate and analyze these patterns?

Drawing on a rich data set from the HarvardX-MITx Person-Course records [37], this study integrates algorithmic enhancements—including temporal extensions in K-Means and adaptive epsilon thresholds in DBSCAN—alongside engagement-aware definitions of low-activity learners. The aim is to uncover actionable insights into learner behaviors that contribute to learning gaps, laying the groundwork for personalized feedback systems and adaptive course designs that can help mitigate these gaps and improve learning outcomes in large-scale online education.

2 Related Work

Clustering has emerged as a key technique in educational data mining for uncovering latent patterns in learner behavior, particularly in large-scale environments like MOOCs. Traditional algorithms such as K-Means have been widely adopted to group learners based on engagement metrics, including video completions, forum interactions, and quiz attempts [3]. For example, [16] applied K-Means to cluster learners by virtual engagement and e-learning infrastructure, demonstrating its utility in identifying distinct engagement categories. However, the application of clustering in MOOCs is not without challenges. K-Means often requires prior specification of the number of clusters and assumes spherical cluster shapes, which may not reflect the heterogeneous engagement patterns typical in MOOCs [1]. [40] applied Ward's hierarchical clustering to initialize K-Means on 1,489 MOOC learners, yielding "active", "passive", and "bystander" groups. [41] combined hierarchical clustering and K-Means to classify 175 learners into four styles—*high/low input–output*. DBSCAN, a density-based algorithm, offers an alternative by identifying clusters of arbitrary shape and handling noise more effectively [18]. [15] revisited DBSCAN to highlight its advantages in processing noisy, irregular data—a characteristic well suited to MOOCs, where learner participation is often sporadic. Despite their effectiveness, these methods typically require adaptations to fully capture the complexity of MOOC data. For instance, [9] employed temporal clustering to analyze weekly participation trends, and revealed that engagement often fluctuates around assessment deadlines. This points to the importance of integrating temporal dynamics into clustering frameworks to better capture learner trajectories over time.

Understanding *engagement patterns* is critical for diagnosing learning gaps in MOOCs. Multiple studies have demonstrated that consistent participation across key activities—such as watching videos, completing quizzes, and posting in forums—is correlated with higher course completion rates [17,19]. Conversely, irregular or minimal engagement often signals emerging learning gaps [25]. Temporal analyses add further depth to this understanding. Research by [20] employed time-series clustering to reveal that drops in engagement frequently coincide with complex course topics or post-assessment periods, which are pivotal moments for intervention. Meanwhile, studies have also highlighted that engagement tends to peak during assessments, leaving substantial gaps during less structured course phases [3]. However, engagement is not uniform across learners due to diverse educational, cultural, and professional backgrounds [4]. This heterogeneity complicates efforts to identify universal engagement-risk patterns, underscoring the need for clustering methods that can flexibly accommodate diverse learner profiles.

Although substantial progress has been made in applying clustering and engagement analytics to MOOCs, several gaps persist. First, many clustering applications do not incorporate temporal or engagement-weighted features, limiting their ability to capture asynchronous participation patterns intrinsic to MOOCs [9,10]. Second, few studies explicitly redefine low-engagement points as indicators of learning gaps rather than treating them as mere noise [10]. Moreover, the integration of clustering with causal interpretability remains limited. Existing studies often stop at descriptive segmentation, without exploring the key factors driving observed learning gaps [23]. This study addresses these gaps by extending clustering algorithms with temporal and engagement-aware adaptations and by employing interpretative metrics to validate patterns indicative of learning gaps.

3 Methodology

3.1 Dataset, Preprocessing and Feature Engineering

The analysis utilizes the HarvardX-MITx Person-Course dataset [37], which provides comprehensive records of learner interactions across multiple MOOC offerings on the edX platform. Each record represents an individual's activity in a single course, containing demographic details, engagement metrics, and outcomes such as final grades. MOOC datasets are inherently sparse and noisy due to asynchronous participation and high dropout rates [2,3]. Consistent with practices in MOOC learning analytics, this study explicitly treats missing engagement values as indicators of non-participation [37]. Engagement features such as nevents, ndays_act, nplay_video, and nchapters were zero-imputed to reflect structural non-engagement, while final grades (grade) were also zero-imputed to indicate absence of course completion credit. For partially missing data, median imputation was applied to numeric features and mode imputation to categorical variables [7,34,39]. To enrich the dataset, several derived features were constructed:

- **Engagement intensity:** A smoothed measure capturing activity density, calculated as $(nevents+1)/(ndays\_act+2)$, inspired by smoothing techniques for handling zero-inflated engagement [36].
- **Quiz-to-video ratio:** Reflects the balance of assessment-driven engagement, computed as $(nchapters + 1)/(nplay\_video + 2)$.
- **Learning duration:** Number of days between course registration and last recorded interaction.
- **Weekly engagement:** Time-series data generated for each learner, mapping their distribution of activity across the official course timeline, enabling temporal pattern analysis [9].

All continuous features were standardized using z-score normalization to ensure comparability in subsequent clustering [39]. Given the asynchronous and often fluctuating participation in MOOCs [20], additional temporal summaries were engineered:

- **Inactive weeks count:** Number of weeks with zero recorded engagement.
- **Engagement drop-off ratio:** Ratio of engagement in the latter half of the course to the first half, indicating mid-course disengagement.
- **Engagement slope and variability:** Derived from linear regression of weekly engagement to quantify trends and participation consistency.

These metrics capture nuanced learner trajectories, addressing the documented limitations of standard clustering techniques in MOOCs, which often overlook temporal dimensions [9].

3.2 Clustering and Outlier Detection

To prioritize engagement and demographic features *most* predictive of learning gaps, logistic regression was employed to learn weights for identifying underperforming learners (defined as achieving a final grade $<= 0.5$). The absolute coefficients from this model were used to scale each feature, generating a set of hybrid-weighted features. Weighted features were then clustered using K-Means, with k=5 chosen to balance granularity and interpretability. This extension of traditional K-Means allows domain-informed weighting to guide the clustering toward engagement patterns more closely linked with learning gaps. To complement K-Means and explore the data's structure without imposing fixed cluster shapes or counts, OPTICS was employed. As an extension of DBSCAN, OPTICS identifies clusters across varying density levels, mitigating the sensitivity of DBSCAN to a single global ϵ parameter and allowing the discovery of clusters with diverse participation densities [15]. This is particularly relevant for MOOCs, where engagement heterogeneity often leads to clusters of varying sizes and densities [10].

To further enhance the detection of atypical engagement patterns potentially indicative of critical learning gaps, the study incorporated a Local Outlier Factor (LOF) analysis. LOF is an unsupervised anomaly detection algorithm that identifies data points exhibiting substantially lower local density compared to their neighbors, effectively capturing learners with irregular or exceptionally low participation trajectories [38]. Such hybrid approach complements clustering by isolating individual learners whose engagement behaviors deviate sharply even within established clusters. In the context of MOOCs, where engagement is already highly heterogeneous LOF enables the identification of subtle yet meaningful outlier profiles that might otherwise be obscured in global cluster summaries. Following standard practice, LOF was applied to the same weighted feature set used in clustering. Detected outliers were analyzed to explore their relationship with poor learning outcomes, reinforcing the linkage between engagement anomalies and learning gaps.

Cluster quality was assessed using multiple internal validation metrics: *Silhouette score*, *Calinski-Harabasz index* and *Davies-Bouldin index* [42]. Additionally, the proportion of low-performing learners in each cluster was analyzed to validate that identified groups meaningfully captured engagement patterns linked to learning gaps.

3.3 Identifying Specific Engagement Patterns Indicative of Learning Gaps

To address the second research question (*RQ2*), this study builds directly on the clustering framework established for *RQ1*, with additional methodological steps focused on temporal engagement profiling and pattern interpretation. Recognizing that learner engagement often fluctuates at different stages of a MOOC [9,20], the course timeline was segmented into three distinct phases:

- **Early phase:** First third of the course weeks, typically reflecting initial exploratory engagement.
- **Mid phase:** Middle third, often encompassing core assessments and sustained learning.
- **Late phase:** Final third, where engagement may decline or stabilize depending on learner persistence.

For each learner, weekly engagement was mapped to the defined course phases, enabling aggregate analyses of both engagement intensity and participation distribution over time. The same hybrid-weighted feature approach from RQ1 was applied, integrating temporal engagement summaries—such as slope, variability, inactive weeks, and drop-off ratios—alongside overall and demographic features. To interpret and validate the identified engagement patterns, three complementary analyses were conducted. First, *cluster profiling* based on average weekly engagement trajectories revealed distinct behavioral archetypes, including learners who began with high activity but quickly disengaged, as well as those with consistently low or stable participation. Second, *engagement intensity* and the *proportion of active learners* were computed separately for the early, mid, and late course phases, shedding light on when disengagement was most likely to occur. Finally, *performance overlays* were applied to each cluster by calculating the proportion of learners with low final grades (less than 0.5), confirming that specific engagement trajectories were strongly associated with learning gaps.

This multi-step approach, integrating temporal segmentation, hybrid clustering, and detailed cluster-wise profiling, provides a nuanced understanding of how specific engagement behaviors throughout a MOOC correlate with underperformance and potential learning gaps.

4 Results

Initial inspection of the HarvardX-MITx dataset revealed high rates of structural non-engagement, with 90.16% of learners exhibiting no recorded video interactions and 62.04% showing no activity in key course events such as problem sets and chapters. After explicitly addressing these structural patterns through zero-imputation and feature engineering, the dataset supported robust clustering and pattern detection.

Key engineered features included *engagement intensity*, measured as the smoothed ratio of events to days active, with standardized values ranging from −0.3538 to 0.3072, effectively captured the spectrum from sparse to dense participation. The quiz-to-video ratio revealed low balanced values for many learners, with most ratios below 1, suggesting limited assessment follow-up relative to content viewed. Additionally, learning durations were long in many cases—for example, spanning up to 334 days—indicating that some learners registered well before or remained active long after the formal course period. Correlation analysis (via heatmap) confirmed only modest associations among most features, supporting the need for multivariate clustering.

The analysis of the 21-week course duration (Table 1) revealed that while total engagement remained substantial even in later stages, the *number of learners actively participating dropped by more than 50%* from early to late phases. Additional pre- and post-course engagement analysis showed that many learners had periods of engagement extending outside the official course window. For example Learner 3 recorded 30 events prior to the course start date, while several

Table 1. Distribution of Events and Learners Across Course Phases

Phase	Total Events	Active Learners
Early	1,161,473 events	50,198 learners
Mid	857,626 events	28,646 learners
Late	1,014,960 events	20,501 learners

Table 2. Percentage of Low-Performing Learners per Cluster

Cluster	% Low Performance
0	**99.06**
1	**100.00**
2	68.21
3	20.41
4	25.00

Table 3. Feature Importance Scores for Predicting Learning Gaps

Feature	Importance
quiz_to_video_ratio	**1.4787**
age	**0.5450**
inactive_weeks_count	0.3065
education_encoded	0.3023
gender_encoded	0.2044
engagement_variability	0.1427
engagement_slope	0.1389
engagement_dropoff_ratio	0.0647
engagement_intensity	0.0739
nforum_posts	0.0308

learners maintained event traces post-course but often at negligible proportions of their total engagement. This suggests exploratory or sporadic behaviors are not strictly tied to course pacing. Moreover, after weekly engagement normalized, we observed highly concentrated activity in the first few weeks followed by long periods of inactivity (normalized values dropping significantly after week 3), and, more consistent learners showed smaller fluctuations around the mean throughout all 21 weeks. The logistic regression model used to weight features for clustering underperformance (Table 3) revealed that assessment-to-content engagement balance, demographic factors, and temporal irregularities were most predictive of learning gaps. Using these weights, K-Means produced *Silhouette score:* 0.983, indicating highly well-separated clusters, *Calinski-Harabasz index:* 162,775.47, and, *Davies-Bouldin index:* 0.333, where lower values suggest excellent separation. Additionaly, cluster breakdown by underperformance (grade <= 0.5, see Table 2) identified *Clusters 0 and 1 as critical profiles*, strongly associated with disengagement and learning gaps. Moreover, weekly engagement trajectories for these clusters (Fig. 2) confirmed that *Cluster 0 and 1* show minimal sustained activity, and quick drop-offs after initial weeks, and *Cluster 3 and 4* are more moderate and show stable engagement patterns (Fig. 1).

OPTICS produced a highly granular structure: a total of 227 microclusters were identified, with an additional 1,079 learners labeled as outliers or noise. The largest clusters included: *Cluster 0*: 10,314 learners, *Cluster 145*: 1,684 learners, and *Cluster 167*: 1,655 learners. Many clusters contained fewer than 30 learners, capturing highly localized behavioral patterns that K-Means did not isolate. The reachability plot (Fig. 2) visually confirmed dense cores interspersed with wide areas of scattered engagement, indicating significant heterogeneity in learning behaviors.

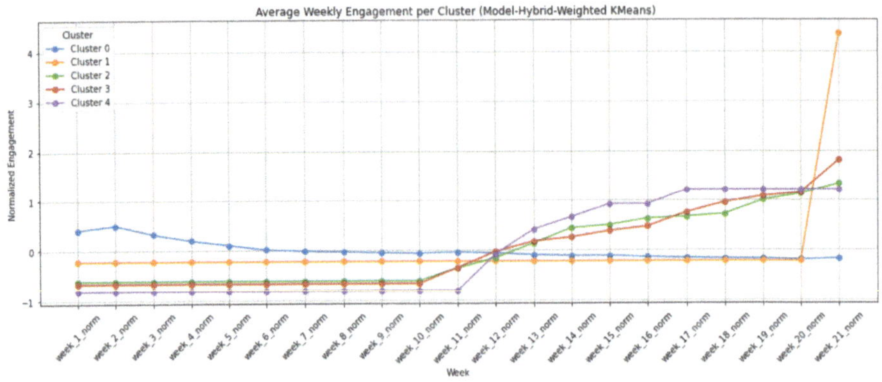

Fig. 1. Average Weekly Engagement per Cluster

To complement the clustering analyses, a *Local Outlier Factor (LOF)* approach was employed on the same hybridweighted feature set. LOF computes the local density deviation of each learner with respect to its neighbors, effectively isolating learners whose engagement patterns were anomalously low or inconsistent even within similar local contexts. The LOF analysis flagged a significant share of learners in Clusters 0 and 1 as *strong local outliers*, consistent with these clusters' extremely high proportions of underperformance (99–100%). Visual inspections (via scatter plots colored by LOF scores, not shown here) further confirmed that learners with high LOF scores corresponded to profiles with minimal participation across weeks and substantial inactive periods. This hybrid anomaly detection reinforced the evidence from clustering that sharp deviations from local engagement norms, as detected by LOF, are tightly linked to learning gaps in MOOCs.

5 Discussion

This study explored how clustering algorithms and anomaly detection techniques could be adapted to identify patterns of learning gaps in MOOCs, using rich engagement data. The findings highlight several important insights. First, the application of hybrid-weighted K-Means revealed clusters with extremely high proportions of underperforming learners (99–100%), underscoring the effectiveness of incorporating supervised feature weights—especially on metrics like quiz-to-video ratio, age, and inactive weeks count—in steering clustering outcomes toward identifying at-risk profiles. Similarly, OPTICS offered a complementary perspective by exposing fine-grained density structures and small, localized learner cohorts, illustrating the heterogeneous nature of MOOC engagement behaviors. The LOF analysis added a local perspective, identifying about 5% of learners as anomalies with notably inconsistent engagement—either unusually sparse or erratically high. Although these outliers had slightly higher mean grades, most still did not complete the course, highlighting that sharp deviations

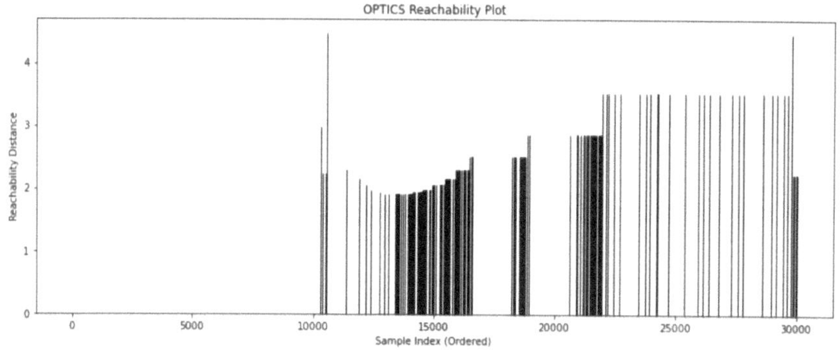

Fig. 2. OPTICS Reachability Plot

from local engagement norms are also tied to learning gaps. Moreover, temporal analyses demonstrated that substantial learner drop-off occurred between early and late phases of the course, despite sustained aggregate event counts. This highlights a core challenge in MOOCs: maintaining consistent participation across the entire course duration.

These contributions, however, should be interpreted in light of certain limitations. The clustering outcomes were assessed using internal validation metrics (silhouette, Calinski-Harabasz, Davies-Bouldin) and grade overlays. However, without an independent test set or alternative course data for cross-validation, generalizability to other MOOCs remains an open question. Future work could validate these methods on additional datasets from different disciplines or platforms to confirm the robustness of the identified patterns. In addition, several feature engineering choices may have constrained the analysis. Metrics such as engagement intensity and quiz-to-video ratio were computed as static aggregates, potentially oversimplifying the temporal dynamics of learner behavior. Applying dynamic time-series clustering or sequence mining techniques (e.g., dynamic time warping) could offer a more nuanced view of evolving engagement patterns. Moreover, the feature set lacked certain key learner attributes. Although age, education level, and gender were included, other potentially influential factors—such as prior subject knowledge or language proficiency—were unavailable. Integrating richer demographic or psychometric data could enhance the interpretability and precision of the resulting clusters. Finally, model configuration choices also warrant refinement. LOF contamination thresholds and OPTICS parameters were selected using standard defaults, which may limit detection sensitivity. Employing more rigorous hyperparameter tuning strategies, such as grid search or Bayesian optimization, could improve the robustness and specificity of anomaly detection.

6 Conclusions

This study demonstrated how advanced clustering and anomaly detection techniques can support the identification of engagement patterns linked to learning gaps in MOOCs. By combining hybrid-weighted K-Means, OPTICS, and LOF, the analysis revealed both broad and fine-grained learner profiles that correlate with underperformance. The results showed that clusters with high proportions of low-performing learners were characterized by early disengagement and irregular participation. Temporal analyses also indicated a sharp decline in learner activity between the early and late phases of the course, despite overall engagement volumes remaining stable. These findings highlight the value of integrating supervised feature weighting and temporal indicators into unsupervised learning frameworks.

In practical terms, this research provides MOOC instructors and platform designers with a refined toolkit. Instead of targeting all low-engagement learners, interventions can now be precisely tailored: proactive support for those showing early warning signs of disengagement, strategic content review for those with imbalanced quiz-to-video ratios, and motivational outreach for those exhibiting erratic but potentially salvageable activity. By focusing on the shape and nature of engagement over time, stakeholders can move beyond simplistic metrics and implement support mechanisms that are both timely and contextually appropriate.

Nonetheless, several limitations remain. The study used a single dataset with aggregated features, potentially missing temporal patterns in learner behavior. Future work could employ time-series data, test across diverse courses, and include richer learner attributes (e.g., prior experience, motivation) to improve explanatory power. Better hyperparameter tuning for anomaly detection may also increase sensitivity to subtle engagement shifts, enabling more targeted, data-driven interventions in large-scale online learning.

References

1. Romero, C., Ventura, S.: Educational data mining: a review of the state of the art. IEEE Trans. Syst. Man Cybern. C (Appl. Rev.) **40**(6), 601–618 (2010)
2. Jordan, K.: Massive open online course completion rates revisited: assessment, length and attrition. Int. Rev. Res. Open Distrib. Learn. **16**(3), 341–358 (2015)
3. Khalil, M., Ebner, M.: Clustering patterns of engagement in massive open online courses (MOOCs): the use of learning analytics to reveal student categories. J. Comput. High. Educ. **29**, 114–132 (2017)
4. Liyanagunawardena, T.R., Parslow, P., Williams, S.: Dropout: MOOC participants' perspective (2014)
5. Le Quy, T., Friege, G., Ntoutsi, E.: A review of clustering models in educational data science toward fairness-aware learning. Educ. Data Sci.: Essent. Approaches Tend. 43–94 (2023)
6. ALSaad, F., Alawini, A.: Unsupervised Approach for Modeling Content Structures of MOOCs. Int. Educ. Data Min. Soc. (2020)

7. Al-Shabandar, R., Hussain, A.J., Liatsis, P., Keight, R.: Analyzing learners behavior in MOOCs: an examination of performance and motivation using a data-driven approach. IEEE Access **6**, 73669–73685 (2018)
8. Brightwood, S., Olaoye Godwin, J.O., Ray, R.: AI-based learning analytics: identifying knowledge gaps and learning styles (2024)
9. Boroujeni, M.S., Dillenbourg, P.: Discovery and temporal analysis of MOOC study patterns. J. Learn. Anal. **6**(1), 16–33 (2019)
10. Mistry, V., Pandya, U., Rathwa, A., Kachroo, H., Jivani, A.: AEDBSCAN—adaptive epsilon density-based spatial clustering of applications with noise. In: Panigrahi, C.R., Pati, B., Mohapatra, P., Buyya, R., Li, K.-C. (eds.) Progress in Advanced Computing and Intelligent Engineering. AISC, vol. 1199, pp. 213–226. Springer, Singapore (2021). https://doi.org/10.1007/978-981-15-6353-9_20
11. Yu, J.H.: Integrating actionable analytics into learning design for MOOCs: a design-based research. J. Comput. High. Educ. (2024)
12. Ruipérez-Valiente, J.A., Muñoz-Merino, P.J., Delgado Kloos, C.: Detecting and clustering students by their gamification behavior with badges: a case study in engineering education. Int. J. Eng. Educ. **33**(2-B), 816–830 (2017)
13. Romero, C., Ventura, S.: Educational data mining and learning analytics: an updated survey. Wiley Interdiscip. Rev. Data Min. Knowl. Discov. **10**(3), e1355 (2020)
14. Kuzilek, J., Hlosta, M., Zdrahal, Z.: Open university learning analytics dataset. Sci. Data **4**(1), 1–8 (2017)
15. Schubert, E., Sander, J., Ester, M., Kriegel, H.P., Xu, X.: DBSCAN revisited, revisited: why and how you should (still) use DBSCAN. ACM Trans. Database Syst. **42**(3), 1–21 (2017)
16. Trivedi, S., Patel, N.: Clustering students based on virtual learning engagement, digital skills, and e-learning infrastructure: applications of k-means, dbscan, hierarchical, and affinity propagation clustering. Sage Sci. Rev. Educ. Technol. **3**(1), 1–13 (2020)
17. Gardner, J., Brooks, C.: Student success prediction in MOOCs. User Model. User-Adap. Inter. **28**(2), 127–203 (2018). https://doi.org/10.1007/s11257-018-9203-z
18. Ester, M., Kriegel, H.-P., Sander, J., Xu, X.: A density-based algorithm for discovering clusters in large spatial databases with noise. In: Proceedings of the KDD, vol. 96, no. 34, pp. 226–231 (1996)
19. Kennedy, G., Coffrin, C., De Barba, P., Corrin, L.: Predicting success: how learners' prior knowledge, skills and activities predict MOOC performance. In: Proceedings of the 5th International Conference on Learning Analytics and Knowledge, pp. 136–140 (2015)
20. Peach, R.L., Yaliraki, S.N., Lefevre, D., Barahona, M.: Data-driven unsupervised clustering of online learner behaviour. NPJ Sci. Learn **4**(1), 14 (2019)
21. Moubayed, A., Injadat, M., Shami, A., Lutfiyya, H.: Student engagement level in an e-learning environment: clustering using k-means. Am. J. Dist. Educ. **34**(2), 137–156 (2020)
22. Kostopoulos, G., Panagiotakopoulos, T., Kotsiantis, S., Pierrakeas, C., Kameas, A.: Interpretable models for early prediction of certification in MOOCs: a case study on a MOOC for smart city professionals. IEEE Access **9**, 165881–165891 (2021)
23. Wang, H., Hao, X., Jiao, W., Jia, X.: Causal association analysis algorithm for MOOC learning behavior and learning effect. In: Proceedings of the IEEE 14th International Conference on Dependable, Autonomic and Secure Computing, pp. 202–206 (2016)

24. Zhou, S., Cao, L., Zhang, R., Sun, G.: M-ISFCM: a semi-supervised method for anomaly detection of MOOC learning behavior. In: International Conference on Pioneering Computer Science Engineering Education, pp. 323–336 (2022)
25. Shi, H., Zhou, Y., Dennen, V.P., Hur, J.: From unsuccessful to successful learning: profiling behavior patterns and student clusters in massive open online courses. Educ. Inf. Technol. **29**(5), 5509–5540 (2024)
26. Dudek, A.: Silhouette index as clustering evaluation tool. Classif. Data Anal.: Theory Appl. **28**, 19–33 (2020)
27. Nelli, F.: Data visualization with matplotlib and seaborn. In: Python Data Analysis: With Pandas, NumPy, and Matplotlib, pp. 183–257 (2023)
28. Shayan, P., van Zaanen, M.: Predicting student performance from their behavior in learning management systems. Int. J. Inf. Educ. Technol. **9**(5), 337–341 (2019)
29. Ren, Z., Rangwala, H., Johri, A.: Predicting performance on MOOC assessments using multi-regression models. arXiv preprint arXiv:1605.02269 (2016)
30. Baker, R.S., Martin, T., Rossi, L.M.: Educational data mining and learning analytics. In: The Wiley Handbook of Cognition and Assessment: Frameworks, Methodologies, and Applications, pp. 379–396 (2016)
31. Agyemang, E.F., et al.: Predicting students' academic performance via machine learning algorithms: an empirical review and practical application. Comput. Eng. Intell. Syst. **15**(1), 86 (2024)
32. Li, Z., et al.: Students' online learning adaptability and their continuous usage intention across different disciplines. Humanit. Soc. Sci. Commun. **10**(1), 1–10 (2023)
33. Halde, R.R.: Application of machine learning algorithms for betterment in education system. In: Proceedings of the International Conference on Automation Control Dynamic Optimization Technology, pp. 1110–1114 (2016)
34. García-Laencina, P.J., Sancho-Gómez, J.-L., Figueiras-Vidal, A.R.: Pattern classification with missing data: a review. Neural Comput. Appl. **19**, 263–282 (2010)
35. Mining, W.I.D.: Data mining: concepts and techniques. Morgan Kaufinann **10**(559–569), 4 (2006)
36. Chen, S.F., Goodman, J.: An empirical study of smoothing techniques for language modeling. Comput. Speech Lang. **13**(4), 359–394 (1999)
37. Ho, A.D., et al.: HarvardX and MITx: two years of open online courses fall 2012-summer 2014 (2015)
38. Breunig, M.M., Kriegel, H.-P., Ng, R.T., Sander, J.: LOF: identifying density-based local outliers. In: Proceedings of the 2000 ACM SIGMOD International Conference on Management of Data, pp. 93–104 (2000)
39. Han, J., Kamber, M., Pei, J.: Data Mining: Concepts and Techniques. Morgan Kaufmann Publishers (2012)
40. Tseng, S.-F., Tsao, Y.-W., Yu, L.-C., Chan, C.-L., Lai, K.R.: Who will pass? Analyzing learner behaviors in MOOCs. Res. Pract. Technol. Enhanc. Learn. **11**(1), 1–11 (2016). https://doi.org/10.1186/s41039-016-0033-5
41. Li, Y.: Evaluation of learning efficiency of massive open online courses learners. Int. J. Emerg. Technol. Learn. **17**(17), 50–61 (2022)
42. Palani, K., Stynes, P., Pathak, P.: Clustering techniques to identify low-engagement student levels. In: Proceedings of the International Conference on Computer Data Mining: Concepts and Technique, vol. 2, pp. 248–257. SCITEPRESS, Setúbal (2021)

Who Pays, Who Reaches? A Data-Driven Analysis of Student Housing Affordability and Accessibility

Kavya Hanumanthu, Stefania Zourlidou(✉), and Frank Hopfgartner

University of Koblenz, Koblenz 56070, Germany
kavyahanumanthu@uni-koblenz.de ,
zourlidou@uni-koblenz.de
https://www.uni-koblenz.de/de/informatik/west

Abstract. We investigate student housing in Koblenz through two lenses, *affordability* (rent, €/m^2, rent-burden) and *accessibility* (distance to campus, commute time, proximity to public transport), by integrating survey microdata with geospatial indicators and density-based clustering. Regression analyses show absolute rent is primarily driven by room size, while €/m^2 is higher for smaller units and, on average, for international students, who also face greater odds of high rent burden. Commute time scales with distance and exhibits no status effect once distance is controlled; however, international students attend campus fewer days per week and are more frequently located in peripheral or weaker-connected areas. Perception evidence aligns with these outcomes: international students prefer to live closer and expect larger benefits from proximity. Findings imply that *who pays* and *who reaches* diverge systematically; policy should moderate small-unit price premiums and prioritize well-connected placements to align preferences with assignments and reduce participation gaps.

Keywords: Student housing · Affordability · Accessibility · Geospatial analysis · Koblenz · data analytics

1 Introduction

Affordable, reachable housing is foundational for students' well-being, participation, and academic performance. In many European university cities, two structural pressures dominate: *affordability* (absolute rent levels, rent per square meter, and rent burden) and *accessibility* (proximity and travel time to campus and essential amenities). This paper adopts a data-driven approach to characterize these pressures and their interaction through a *case study in Koblenz, Germany*, using survey and geospatial data from students enrolled at the *University of Koblenz*. We focus on two questions: *RQ1 (Affordability)*—How affordable is student housing when operationalized via rent levels, rent per m^2, and rent-burden indicators? *RQ2 (Accessibility)*—How accessible is student housing in

M. E. Auer et al. (Eds.): IMCL 2025, LNNS 1954, pp. 201–212, 2026.
https://doi.org/10.1007/978-3-032-24998-2_17

terms of campus proximity, commute time, and reachability via public transport?

Affordability is typically expressed as the relationship between housing costs and resources; practical indicators include rent-to-expenditure shares and rent per m^2 [1–3]. For students, direct income measures are noisy, so rent-share proxies and per-area costs are robust, comparable signals [2,3]. *Accessibility* denotes the ease of reaching opportunities across land-use and transport systems [5,6]. Beyond averages, transport equity requires attention to distributive justice—i.e., whether different groups experience comparable travel times and campus reachability [7,8]. European evidence documents widespread affordability stress among students, with rent often the largest budget component [2,3]. We therefore expect substantial variance in rent per m^2 and rent-burden rates across housing types and locations. On the access side, commute-time and performance links suggest that distance and network structure meaningfully shape students' academic lives [5,9]. Importantly for our context, Koblenz's river-valley geography and bridge-dependent street network can decouple straight-line distance from true travel time; QGIS-based geocoding and mapping help quantify such realities [11].

We integrate survey features (costs, room size, attendance, perceived importance of proximity) with geospatial measures (distance to campus; proximity to public transport hubs). For *RQ1*, we analyze rent distributions, rent per m^2, and rent-burden indicators via regression models. For *RQ2*, we examine distancetime relationships and spatial structure, including density-based clustering to surface residential concentrations and peripheral outliers [14]. Throughout, we interpret results through an equity lens, considering whether international students face systematically worse affordability and accessibility conditions [4,7,8].

Contributions. **(C1, RQ1)** Operational, interpretable affordability indicators (rent, €/m^2, rent-burden) instantiated on a *University of Koblenz* student sample. **(C2, RQ2)** A joint spatialbehavioural accessibility analysis (distance to campus, commute time, proximity to PT hubs, attendance) tailored to the *Koblenz* urban network. **(C3)** Density-aware exploration (DBSCAN) to surface central clusters and peripheral pockets. **(C4)** Evidence of group disparities (international vs. domestic) with policy implications for pricing of small units and placement in connected catchments [4,7,8].

2 Related Work

Research on student housing consistently highlights two structural dimensions that shape outcomes: *affordability* and *accessibility*. Affordability is commonly operationalized through rent-to-resource relationships (e.g., expenditure shares) and price-per-area indicators, reflecting both absolute and normalized cost burdens [1–3]. European and German evidence indicates that rent is often the largest budget component for students, with sizable variation by dwelling type and city, and with international students frequently facing higher barriers in the housing market [2–4]. Within housing studies, affordability is also linked to subjective well-being and persistence, as cost pressures compound other challenges faced by low-income and mobile populations [17–19]. In university-town contexts, market segmentation (dormitories, shared flats, studios) and shifting supply conditions can amplify these pressures, producing stratified rent-per-m^2 patterns across neighborhoods [20–22]. In what follows we instantiate these ideas in a river-valley university city—Koblenz, Germany—where the RhineMoselle confluence, bridge dependencies, and a historic/touristic core shape both affordability patterns and effective accessibility.

Accessibility refers to the ease of reaching opportunities and has long been framed at the intersection of land use and transport systems [5,6]. Contemporary perspectives emphasize not only average performance but also *distributive justice*, asking whether different groups experience comparable travel times and campus reachability [7]. In higher education, access to campus resources (libraries, labs, advising) is temporally and spatially structured; thus, small differences in distance and network connectivity can cumulate into participation gaps, particularly for students reliant on public transport [8]. Empirical studies connect commute time to academic performance and attendance, suggesting that location and network impedance are consequential for day-to-day engagement [9]. These strands collectively motivate operational metrics for *RQ2* that go beyond straight-line distance, incorporating network-based travel time and proximity to everyday amenities.

Methodologically, accessibility measurement has benefited from open-source geospatial tooling and reproducible network analytics. OSMnx provides programmatic acquisition and analysis of street networks, enabling more realistic travel-time and centrality computations than Euclidean proxies [10]. GIS platforms (e.g., QGIS) facilitate the integration of survey microdata with network- and point-of-interest layers, supporting the derivation and visualization of distance/time indicators to campus and essential services [11,16]. In student-housing applications, these tools allow researchers to (i) compute multimodal access metrics at residential coordinates, (ii) map spatial heterogeneity in campus reachability, and (iii) examine co-location (or mismatch) between affordable stock and high-access areas. This toolbox underpins data-driven evaluations of *RQ2* at city scale.

Affordability analytics for students typically start with descriptive distributions of rent and rent-per-area, then compare subgroups and dwelling types, often acknowledging that income measurement is noisy in student populations [1,2]. Policy-facing work in Germany further quantifies rent shares and the protective role of institutional housing offers (e.g., Studierendenwerk), though coverage constraints can push many students to the private market where costs and variance are higher [3]. Equity-oriented reports document that international students face compounding barriers (language, documentation, timing), which can translate into both higher effective search costs and less favorable matches in location and price, reinforcing affordability disadvantages [4]. A growing literature connects affordability and accessibility rather than treating them in isolation. In compact or transit-rich settings, proximity can partly offset higher nominal rents (via lower commuting costs and time), while peripheral affordability gains may erode under network bottlenecks or limited service frequency [6,7]. In river-valley geographies and bridge-dependent networks (a common pattern in smaller European cities), shortest-path travel times can be highly asymmetric relative to Euclidean distance, making network-aware measures essential for valid comparisons across neighborhoods [10,16]. For students with rigid course schedules and limited modal flexibility, these network effects can interact with affordability constraints to shape attendance patterns and perceived access [8,9].

Across these strands, three design principles recur and inform our analytic choices for our research questions. *First*, indicators should be interpretable: rent levels, rent-per-area, and rent-burden flags for affordability; network travel time and amenity proximity for accessibility [1,6]. *Second*, methods should reflect how students actually move through cities: network-based measures and city-specific constraints rather than planar distances [10,11]. *Third*, analyses should surface group differences relevant for equity—particularly domestic vs. international status—given evidence of systematic disparities in both cost exposure and effective reachability [4,7,8].

3 Methodology

We operationalize our research questions with interpretable indicators, network-aware measurements, and robust group comparisons. The workflow integrates student survey microdata with geospatial layers and street-network analytics, and employs density-based clustering to reveal spatial structure relevant to access.

3.1 Case Study: Koblenz

Koblenz is a medium-sized German city at the confluence of the Rhine and Moselle. Its historic and touristic core sits along waterfront promenades, while residential districts extend into valley corridors and hillside neighborhoods. Bridge-dependent crossings concentrate flow at a small number of links, so straight-line distance can diverge markedly from actual travel time. Amenity distributions also vary between the compact center and peripheral quarters. These features are directly relevant to our indicators: (i) distance to campus and self-reported commute time (with proximity to public transport hubs) provide a realistic basis for *RQ2*; and (ii) housing submarkets (Studierendenwerk dorms vs. private WG/studios) motivate stratified affordability views for *RQ1*. The survey sample comprises students enrolled at the University of Koblenz; residential coordinates are privacy-aggregated as described below.

3.2 Dataset, Preprocessing, and Feature Engineering

We combine (i) a student-housing survey capturing dwelling characteristics (accommodation type, rent, floor area), demographics (domestic vs. international status), and behaviours (days on campus per week, perceived importance of proximity); (ii) geospatial layers for campus coordinates, public-transport stops, and everyday amenities (supermarkets, healthcare, transit); and (iii) a city street-network extracted from OpenStreetMap for network-based distances and travel times [10,11]. Currency (all monetary fields such as rents and budgets) and area units are standardized. Potential outliers were assessed via boxplots and residual diagnostics; as a robustness check, models were re-estimated after excluding extreme observations, with conclusions unchanged. Responses were screened for completeness and consistency; common placeholders (e.g., "none," "nil," "keine") were standardized and, depending on context, coded as zeros or with the modal category for categorical items. English/German responses were harmonized to a common English coding and merged into one dataset. To protect privacy, respondents provided only street name and postal code (no full addresses). Entries marked "private" were removed; street names were normalized, concatenated with the postal code and "Koblenz, Germany," and geocoded to *approximate* coordinates for spatial analysis (campus/amenity proximity), balancing utility and anonymity.

We compute:

$$\text{RentPerArea} = \frac{\text{Rent}}{\text{Area}}\ [€/\text{m}^2], \qquad \text{RentBurden} = \mathbb{1}\{\text{Rent} > 0.5 \times \text{Monthly Budget}\}.$$

following the survey instrument and affordability operationalizations in the study. *RentPerArea* captures price normalised by space, while *RentBurden* reflects lived financial strain.

Objective accessibility is measured using (i) geographic *distance to campus* computed from geocoded residence locations, (ii) *self-reported commute time* (minutes), and (iii) *proximity to public transport hubs*. Behavioural participation is captured by *days per week on campus*.

3.3 Analytic Strategy

We analyse affordability with the five regression specifications reported in Table 1: (1) OLS for absolute rent; (2)–(4) OLS for €/m^2 with progressive controls (student status, room size, geography); and (5) a logistic model for rent burden ($> 50\%$). Descriptive summaries (medians, IQR) accompany the models, and group differences (domestic vs. international) are inferred from the corresponding regression coefficients rather than from separate hypothesis tests.

Accessibility is examined with the four regression models in Table 2: (1) distance to campus explained by commute time and status; (2) commute time explained by distance and status; (3) the same restricted to bus users; and (4) attendance (days/week on campus) explained by status and distance. In addition, we apply DBSCAN clustering on residential coordinates to identify central clusters and peripheral outliers; the neighbourhood radius is set to $\varepsilon = 1700$ m, chosen via the k-distance elbow, following the study's spatial analysis [14]. Resulting clusters are mapped to describe spatial structure.

Relations between affordability and accessibility are synthesised qualitatively by juxtaposing the regression results (Tables 1–2) with the spatial clustering and maps, highlighting where cost exposure and peripheral location co-occur.

Table 1. Regression models for affordability.

Model	Outcome	Adj.R^2/Pseudo	Key predictors	Key finding
1	Rent (€)	0.449	Room size	+€8.6 per m^2; status n.s.
2	€/m^2	0.045	Status	Intl +€3.62/m^2
3	€/m^2	0.298	Room size	Smaller rooms costlier; status n.s.
4	€/m^2	0.289	Size, Status, Region	Size sig.; Intl effect returns; region n.s.
5	Rent burden	–	Status (logit)	Intl higher odds of $> 50\%$

All geocoding and cartographic outputs were produced and validated in QGIS; statistical models and clustering follow the documented settings used in the study's analysis pipeline [11,14].

4 Results

We present findings for our research questions using the indicators and models described above. Figures referenced here correspond to exported plots that replicate the established outputs.

4.1 Affordability

Regression models (Table 1). Model 1 (absolute rent) identifies room size as the dominant predictor (each additional m^2 adds approximately €8.6); accommodation type and student status are not significant. Model 2 (€ /m^2) shows international students pay on average about €3.62 more per m^2. Model 3 adds room size: the status effect weakens (n.s.) while smaller rooms are costlier per m^2. Model 4 adds geography: room size remains significant, international status regains significance, and region remains n.s. Model 5 (logit) indicates higher odds of rent burden ($> 50\%$) for international students. Affordability gaps by status are shown in Fig. 1a (€/m^2) and Fig. 1b (share $> 50\%$ burden). These plots replicate the established visualizations.

4.2 Accessibility

Regression models (Table 2). Model 1 validates commute time as a strong predictor of distance to campus; status is nonsignificant (n.s.) Model 2 shows commute time explained chiefly by distance; status n.s. Model 3 (restricted to bus users) reproduces the same pattern. Model 4 relates attendance (days/week) to status and distance: international students attend on average 0.85 fewer days/week, while distance is n.s. DBSCAN on residential coordinates with a neighborhood radius $\varepsilon = 1700\,\text{m}$ yields central clusters and several noise points; outliers disproportionately correspond to longer commutes. The accompanying map corroborates that domestic students concentrate near walk/cycle range, while international students appear more often in peripheral or weaker-connected areas (Figs. 2a, 2b).

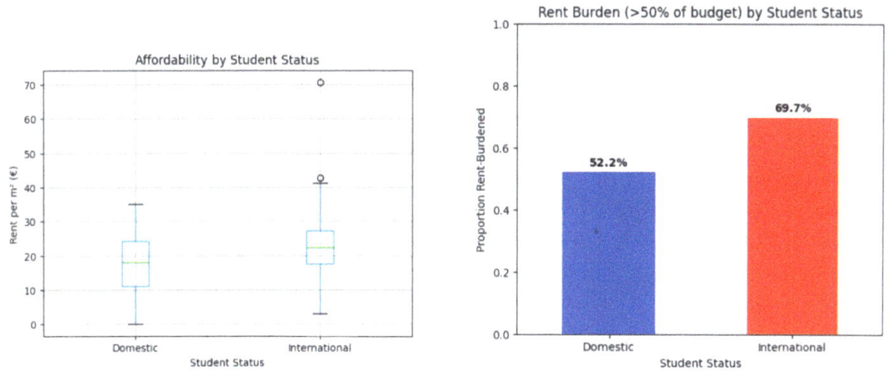

(a) Affordability by status (€/m^2).　　(b) Rent-burden > 50% by status.

Fig. 1. Affordability comparisons by student status: (a) price per area and (b) high rent-burden share.

Table 2. Regression models for accessibility.

Model	Outcome	Adj.R^2	Key predictors	Key finding
1	Distance	0.186	Time, Status	Time $\uparrow \Rightarrow$ Distance $\uparrow$; status n.s.
2	Time (all)	0.315	Distance, Status	Distance $\uparrow \Rightarrow$ Time $\uparrow$; status n.s.
3	Time (bus)	0.373	Distance, Status	Same pattern among bus users
4	Days/week	0.077	Status, Distance	Intl: −0.85 days/week; distance n.s.

Preferred versus actual residential distances show an aspirationoutcome gap: international students more often prefer 1–2 km yet are located 3–5 km or more; domestic students more frequently report that distance "doesn't really matter." Perceived-impact heatmaps show that international students more often state that proximity would improve attendance and academic performance (Figs. 3–4).

4.3 Integrated Findings and Robustnesss

Across outcomes, two patterns are consistent. *Affordability:* absolute rent is chiefly a function of room size, while €/m^2 is higher for smaller rooms and, on average, higher for international students; coarse geography is not significant in the €/m^2 models (Table 1; Fig. 1). *Accessibility:* commute time is explained primarily by distance to campus; once distance is accounted for, time does not differ systematically by status. However, international students attend fewer days per week, and the spatial analysis (DBSCAN) reveals more peripheral/outlier locations among internationals (Table 2; Fig. 2). Perception evidence aligns with these patterns: internationals report stronger benefits from living closer to campus (Fig. 4).

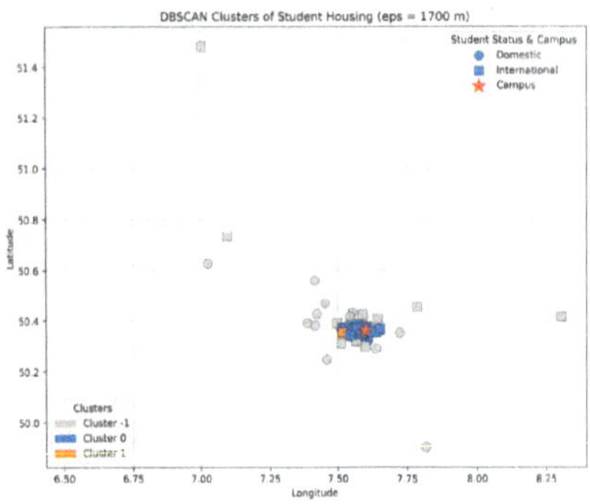

(a) Residential clusters (DBSCAN, ε = 1700 m) with outliers.

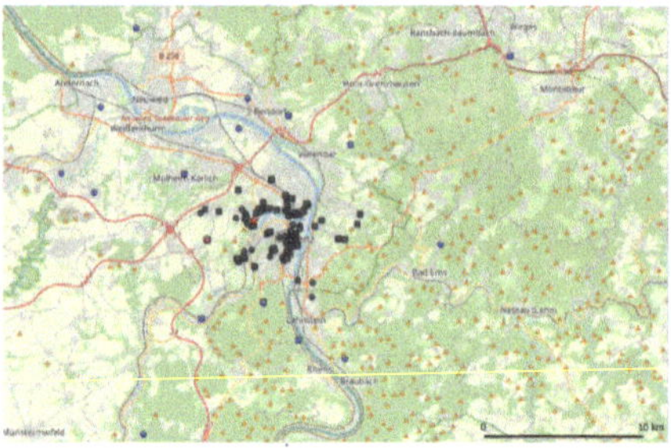

(b) Residence clusters: gray circles (domestic residences), gray and red rectangulars (international residences) and campus (red star), on an OSM basemap.

Fig. 2. Spatial context and clustering of student residences in Koblenz.

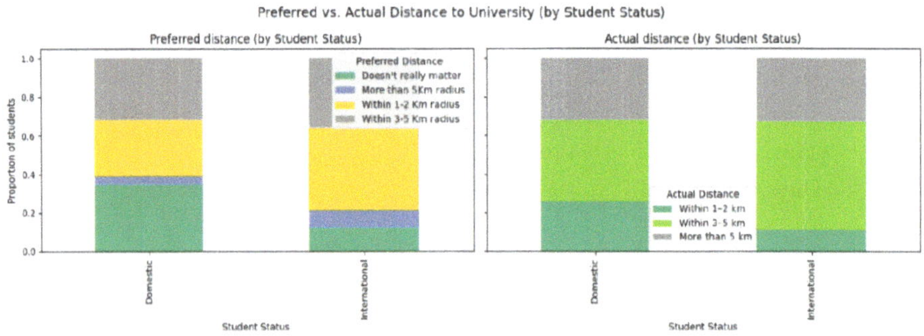

Fig. 3. Preferred vs. actual distance to campus by student status.

Findings are stable under several checks that mirror the analysis pipeline: (i) re-estimating models after excluding extreme observations (identified by boxplots and residual diagnostics) leaves coefficients' signs and substantive conclusions unchanged; (ii) restricting the timedistance model to bus users reproduces the same distancetime relationship and lack of a status effect; (iii) varying the DBSCAN radius in a reasonable neighbourhood around the chosen value (ε = 1700 m) preserves the core clustering pattern—central aggregates plus peripheral noise points with a higher share of internationals; and (iv) using alternative aggregation views (by dwelling type for €/m^2 and by distance bands for time/attendance) yields the same qualitative gaps. Taken together, the evidence shows that **who pays** more and **who reaches** less—especially among international students—reflects durable structural pricing and spatial placement patterns in Koblenz, rather than quirks of any single model.

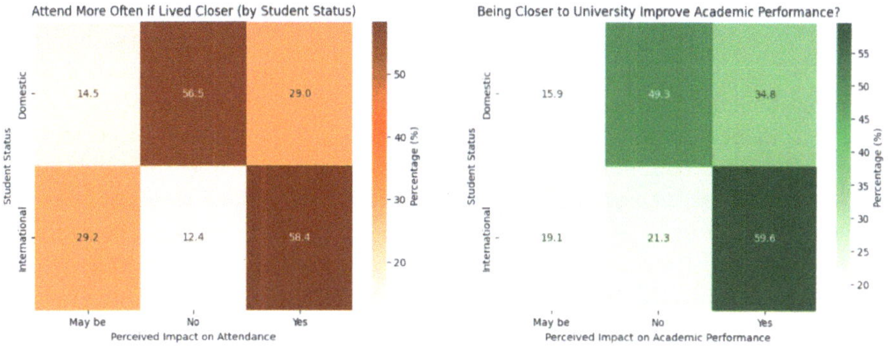

(a) Perceived impact of proximity on *attendance* by status.

(b) Perceived impact of proximity on *academic performance* by status.

Fig. 4. Perceived benefits of living closer to campus by student status: (a) attendance and (b) academic performance.

5 Discussion

This study examined two foundational dimensions of student housing, *affordability* and *accessibility*, through a case study in Koblenz based on student survey microdata and geospatial indicators. We interpret the empirical patterns, reflect on intergroup differences, and draw implications for policy and practice.

Interpreting Affordability (RQ1). Three results are robust across specifications in Table 1. First, *room size* is the dominant structural driver of absolute rent (Model 1: ≈€8.6 per additional m^2). Second, *international status* is associated with higher €/m^2 (Model 2: ≈€3.62), an effect that attenuates once size is controlled (Model 3) but reappears when geography is added (Model 4). Third, international students face higher odds of *rent burden* ($>$ 50% of budget; Model 5). Together, these findings indicate that affordability inequities are primarily *structural* (small rooms command a price premium per m^2) and *demographic* (status-linked penalties), whereas coarse geography is not a significant determinant of €/m^2. Figures 1(ab) visualize these gaps. Two inferences follow. (i) The attenuation and re-emergence of the status effect suggest that internationals' concentration in smaller units explains part—but not all—of the €/m^2 gap. (ii) Higher rent-burden odds indicate that comparable nominal rents translate into greater financial strain for international students, consistent with budget constraints captured in the survey.

Interpreting Accessibility (RQ2). Objective patterns behave as expected: *distance* and *commute time* move together, and after accounting for distance there is no systematic status difference in time (Models 13, Table 2). Behaviourally, however, *attendance* differs: international students attend ≈0.85 fewer days per week (Model 4). The spatial analysis complements these averages: DBSCAN with ε = 1700m identifies central clusters and peripheral outliers; international students are overrepresented among the *noise* points (Fig. 2a), and the basemap indicates more frequent residence in weaker-connected or peripheral areas (Fig. 2b). Perception evidence is aligned: internationals more often *prefer* to live 1–2 km from campus yet are located 35 km+, and they report stronger perceived benefits from proximity for attendance and performance (Figs. 34).

The evidence points to a pragmatic frontier: central areas tend to be costlier yet yield shorter commutes; peripheral areas may be cheaper but risk longer and more variable access. In a river-valley city with bridge dependencies like Koblenz, the trade-off is not smooth—micro-geographic constraints can create pockets that are both relatively costly and operationally inconvenient. Policy should therefore avoid assuming a simple rentdistance gradient and prioritize *effective reachability* alongside price. Moreover, the results clarify *who pays* and *who reaches*. International students *pay* more per m^2 and are more often rent-burdened; they also *reach* campus less often (fewer attendance days) and are more frequently sited at the periphery. These intergroup gaps persist across robustness checks and are visible in both outcomes and perceptions, suggesting durable structural pricing and spatial placement patterns rather than artifacts of a particular model.

Policy implications are fourfold. First, target small-unit premiums: because €/m^2 penalties concentrate in small rooms—and international students are over-represented in such stock—expanding affordable small-format options (or moderating per-m^2 pricing in student-targeted units) addresses a key structural driver. Second, prioritise connected placements: focusing allocations and new stock within walk/bike/transit catchments of campus can mitigate attendance impacts without relying solely on nominal distance reductions; the maps and clusters indicate where connectivity improvements matter most. Third, align preferences and placements: preferencerealisation mismatches (desired 1–2 km vs. realised 3–5 km+) and stronger perceived benefits from proximity among internationals argue for earlier housingadmissions coordination, transparent guidance on locations, and targeted advisory support during arrival periods. Fourth, support participation, not just proximity: since attendance differs by status even when distance is controlled, complementary measures—such as schedule design that reduces mandatory travel frequency, mobility subsidies timed to semester peaks, or on-campus "clustering days" for services—can help close participation gaps.

6 Conclusions

We studied student housing in Koblenz through two lenses—*affordability* (RQ1) and *accessibility* (RQ2)—combining survey microdata with geospatial analysis. The pipeline yields interpretable indicators and links location patterns to behavioural participation.

For affordability, room size is the main driver of absolute rent, while €/m^2 is higher in smaller units and, on average, higher for international students, who also face greater odds of high rent burden; coarse geography does not explain €/m^2 differences. For accessibility, commute time scales with distance and shows no status effect once distance is controlled, yet international students attend fewer days per week. Clustering and maps indicate they more often live in peripheral or less-connected areas, and perceptions reveal stronger benefits from living closer.

In Koblenz, *who pays* and *who reaches* differ systematically: affordability gaps are chiefly structural and demographic, while access gaps arise from where students—especially internationals—reside within the network and how that siting shapes attendance. Policies should address small-unit premiums and prioritise placements in connected catchments; aligning stated preferences with actual assignments and supporting participation (not only proximity) can reduce gaps.

References

1. Hulchanski, J.D.: The concept of housing affordability: six contemporary uses of the housing expenditure-to-income ratio. Hous. Stud. **10**(4), 471–491 (1995)
2. European Commission, DG ECFIN: Housing market developments in the euro area—Focus on housing affordability. Publications Office of the European Union (2022)
3. Deutsches Studentenwerk: 21. Sozialerhebung: Wirtschaftliche und soziale Lage der Studierenden in Deutschland 2021 (2022)
4. Pineda, J., Kercher, J., Falk, S., Thies, T., Yildirim, H.H., Zimmermann, J.: Accompanying international students in Germany to academic success. DAAD—German Academic Exchange Service (2022)
5. Hansen, W.G.: How accessibility shapes land use. J. Am. Inst. Plann. **25**(2), 73–76 (1959)
6. Geurs, K.T., van Wee, B.: Accessibility evaluation of land-use and transport strategies: review and research directions. J. Transp. Geogr. **12**(2), 127–140 (2004)
7. Pereira, R.H., Schwanen, T., Banister, D.: Distributive justice and equity in transportation. Transp. Rev. **37**(2), 170–191 (2017)
8. Shields, R.: Student mobility and access to higher education: implications for equity. High. Educ. **78**(5), 761–777 (2019)
9. Kobus, M.B.W., van Oort, F., Rietveld, P.: Student commute time, university presence and academic achievement. Reg. Sci. Urban Econ. **52**, 129–140 (2015)
10. Boeing, G.: OSMnx: new methods for acquiring, constructing, analyzing, and visualizing complex street networks. Comput. Environ. Urban Syst. **65**, 126–139 (2017)
11. QGIS Development Team: QGIS Geographic Information System. Open Source Geospatial Foundation (2023)
12. Ruxton, G.D.: The unequal variance t-test is an underused alternative to Student's t-test and the Mann-Whitney U test. Behav. Ecol. **17**(4), 688–690 (2006)
13. Cohen, J.: Statistical Power Analysis for the Behavioral Sciences. Routledge, 2nd ed. reprint (2013)
14. Ester, M., Kriegel, H.-P., Sander, J., Xu, X.: A density-based algorithm for discovering clusters in large spatial databases with noise. In: Proceedings of KDD, pp. 226–231 (1996)
15. Rousseeuw, P.J.: Silhouettes: a graphical aid to the interpretation and validation of cluster analysis. J. Comput. Appl. Math. **20**, 53–65 (1987)
16. Fotheringham, A.S., Rogerson, P.A. (eds.): The SAGE Handbook of Spatial Analysis. Sage (2008)

17. Evans, G.W., Wells, N.M., Moch, A.: Housing and mental health: a review of the evidence and a methodological and conceptual critique. J. Soc. Issues **59**(3), 475–500 (2003)
18. Clapham, D.: The Meaning of Housing: A Pathways Approach. Policy Press (2005)
19. Collins, F.L.: Researching mobility and emplacement: examining transience and transnationality in international student lives. Area **44**(3), 296–304 (2012)
20. Garmendia, M., Coronado, J.M., Ureña, J.M.: University students sharing flats: when studentification becomes vertical. Urban Stud. **49**(12), 2651–2668 (2012)
21. Hubbard, P.: Geographies of studentification and purpose-built student accommodation: leading separate lives? Environ Plan A **41**(8), 1903–1923 (2009)
22. Kenna, T.: Studentification in Ireland? Analysing the impacts of students and student accommodation on Cork City. Ir. Geogr. **44**(2), 191–213 (2011)

Work-In-Progress: Quality Data Insights Management System

Abhishek S. Joshi(✉) and Kalyan Ram B.

Indxo AI Pvt Ltd, 3/11, Chanduri, II Cross, I Main, Abhayadhama Road, Whitefield, Bangalore, Karnataka 560066, India
{abhishek.joshi,kalyan}@indxo.ai

Abstract. In today's manufacturing landscape, many industries still rely heavily on isolated inspection processes and manual quality checks, resulting in delayed defect detection, high rework rates, and limited traceability. With disconnected production and quality systems, root cause analysis becomes reactive and inefficient, hindering efforts toward achieving consistent, high-quality output and process optimization.

To address this, the paper presents a dual-layered approach to real-time data acquisition and analysis—one layer focused on pro-duction machines (e.g., CNCs, PLCs) capturing key signals like spindle load, feed rate, and tool condition, and another layer dedicated to quality machines (e.g., CMMs, vision systems, digital gauges) gathering inspection data such as dimensional deviation, surface finish and part pass/fail status. By bridging the digital gap between machine performance and part quality, manufacturers can detect early signs of process drift before they turn into non-conformities. This structured integration creates a unified digital thread, shifting quality management from reactive to predictive and proactive control.

By enabling part-wise traceability, Cp/Cpk process capability analysis, and real-time parameter tracking, the system empowers engineers with actionable insights. Advanced dashboards, barcode or RFID-based traceability, and centralized statistical platforms support continuous improvement and early defect prevention. As data becomes the common language across departments, quality teams and production teams can collaborate more effectively to drive targeted interventions. This convergence of production and quality data drives zero-defect manufacturing and forms the core of Industry 4.0-driven quality intelligence setting the stage for autonomous decision-making and smarter manufacturing ecosystems.

Keywords: Real-time data integration · Quality intelligence · CNC–CMM interoperability · Digital traceability · Process capability analysis (Cp/Cpk)

1 Introduction

In the evolving landscape of smart manufacturing, quality assurance can no longer function as a standalone process. Traditional manufacturing setups, where production and inspection systems operated in silos, often resulted in delayed defect detection, inefficient root cause identification and compromised product integrity. Modern practices

M. E. Auer et al. (Eds.): IMCL 2025, LNNS 1954, pp. 213–222, 2026.
https://doi.org/10.1007/978-3-032-24998-2_18

emphasize the importance of real-time data integration between production machines like CNCs, PLCs and quality systems such as CMMs, vision systems and digital gauges. This integration enables dynamic process control, instant quality feedback and swift corrective actions. As a result, manufacturers can ensure compliance, reduce rework and move closer to the goal of zero-defect manufacturing.

1.1 Characteristics

A tightly integrated production-quality architecture is defined by several essential features. Continuous data flow between machines and inspection devices enables synchronized operations and real-time visibility of both process and product parameters. Digital traceability is maintained through barcode or RFID-based part tracking, while standardized protocols such as OPC-UA and MTConnect support uniform data exchange across diverse platforms. Edge analytics allows for faster data processing at the machine level, ensuring timely decision-making. Interoperability between multi-vendor systems eliminates the need for manual interventions and closedloop feedback mechanisms ensure that inspection results can immediately influence production parameters, such as triggering automatic tool offset adjustments.

1.2 Examples

Practical implementations of such systems illustrate their value. For instance, a CNC machine's spindle load and tool wear data can be directly correlated with CMM measurement deviations to identify trends and take timely action. In aerospace manufacturing, inspection data automatically triggers tool corrections, ensuring consistent precision without operator input. In electronics, vision systems detect surface defects and PLC data helps pinpoint the exact production step responsible. Centralized dashboards built using protocols like OPC-UA aggregate real-time insights from various sources, supporting fast decisions and audit readiness. Similarly, RFID-tagged components in medical device manufacturing are tracked through each production and inspection stage, ensuring full regulatory compliance and quality traceability.

2 Objective

The primary objective of the proposed system is to establish a unified and intelligent quality management framework within a smart manufacturing environment. By integrating operational data from CNC machines and PLCs with inspection data from Coordinate Measuring Machines (CMMs), vision-based systems and digital gauges, the architecture facilitates immediate defect detection and corrective action. Furthermore, it ensures end-to-end digital traceability from raw machining to final quality verification thereby improving compliance and audit readiness. Real-time statistical process control, including Cp/Cpk monitoring, empowers engineers to intervene early and maintain process stability. The architecture is inherently scalable and adaptable, supporting various industrial setups and facilitating consistent improvement in firstpass yield through continuous, data-driven decisions.

3 Detailed Methodology

To achieve seamless integration, the methodology adopts a layered and structured data flow mechanism. At the shop-floor level, real-time data is acquired from production assets such as CNCs, PLCs and embedded sensors (e.g., for temperature, vibration). In parallel, quality checkpoints employ digital inspection tools, including CMMs and vision systems, to capture dimensional and surface integrity metrics. These devices communicate with an edge gateway that manages protocol standardization (e.g., Modbus, OPC-UA), performs initial preprocessing and transmits the harmonized data to a central SQL-based data lake. This infrastructure enables unified, contextualized datasets that support real-time monitoring and historical analysis of both production and quality metrics. By synchronizing machine and inspection data, the system enhances traceability and root-cause analysis capabilities.

3.1 System Setup Architecture

The proposed architecture follows a multi-layered model optimized for interoperability and scalability. At the base layer, data is acquired from various machine tools and inspection systems. This raw data is fed into an edge gateway responsible for acquisition, lightweight analytics and secure transmission. The centralized SQL database acts as a single source of truth, storing structured data for consumption by analytics platforms. Power BI is leveraged for real-time visualization, offering dashboards that present key performance indicators (KPIs), part-wise defect metrics and capability indices (Cp/Cpk). These insights enable engineers and quality teams to detect anomalies, analyze process deviations and implement timely corrective actions. The system transforms disconnected quality and production checkpoints into a cohesive decision-support network, thereby enhancing product quality, minimizing rework and driving operational efficiency (Fig. 1).

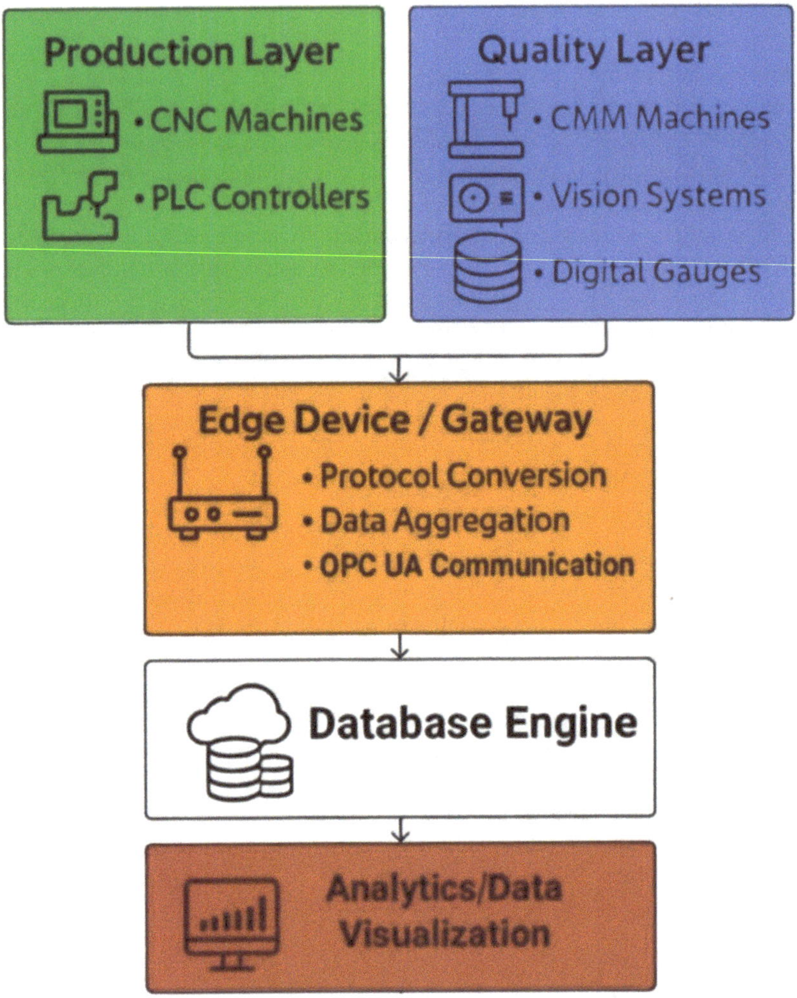

Fig. 1. Architecture Diagram

4 Integrated Operational and Quality Analytics Dashboards

4.1 Dimensional Report

The Dimensional Report dashboard provides a comprehensive statistical visualization of individual measured characteristics obtained from precision measurement systems. For features such as Ø20.000 $\pm$ 0.05, it displays a Boxplot that illustrates the distribution of values, including minimum, first quartile (Q1), median, third quartile (Q3) and maximum helping quickly identify data spread and outliers. A corresponding histogram presents the frequency distribution of these values over selected timeframes, while a trend chart plots the dimensional variation over time to detect gradual drifts, tool wear or process instability. Filters such as Part Name, Characteristic, and Date allow engineers and quality teams to isolate and diagnose variation sources effectively (Fig. 2).

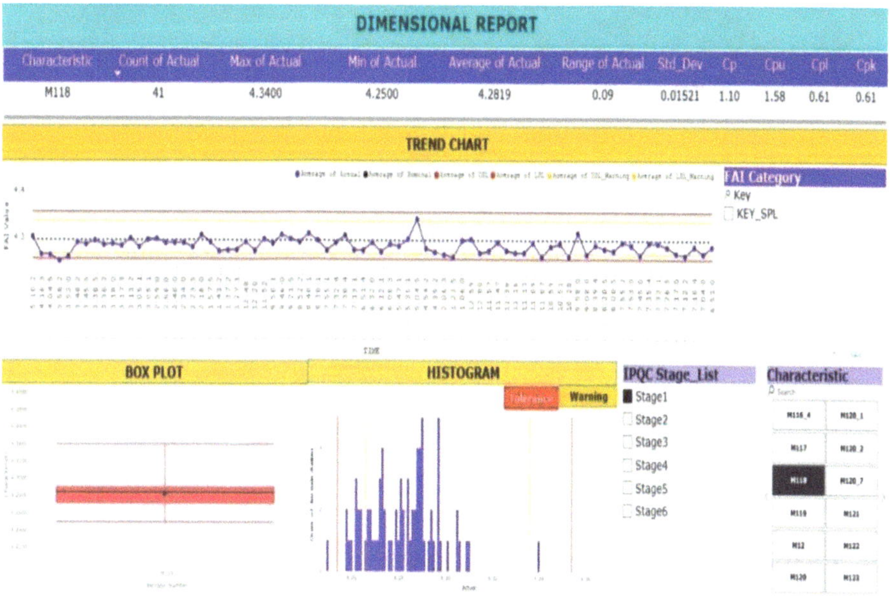

Fig. 2. Stage-Wise Cp/Cpk Distribution for Dimensional Data

4.2 Yield Analysis

The Yield Analysis module visualizes inspection results at the barcode level, categorizing outcomes into PASS or FAIL and mapping them against production shifts and calendar dates. A bar chart, color-coded (e.g., Green for PASS, Red for FAIL), provides instant clarity on product quality trends for Shift A or B on any given day (e.g., 17-Jul25). Each bar represents the total number of scanned barcodes per shift, with accompanying summary labels like Total Scanned and Total Failed. This module supports backward traceability and helps identify shift-specific issues, recurring failure timelines or operator-level performance deviations that can influence overall product conformity (Fig. 3).

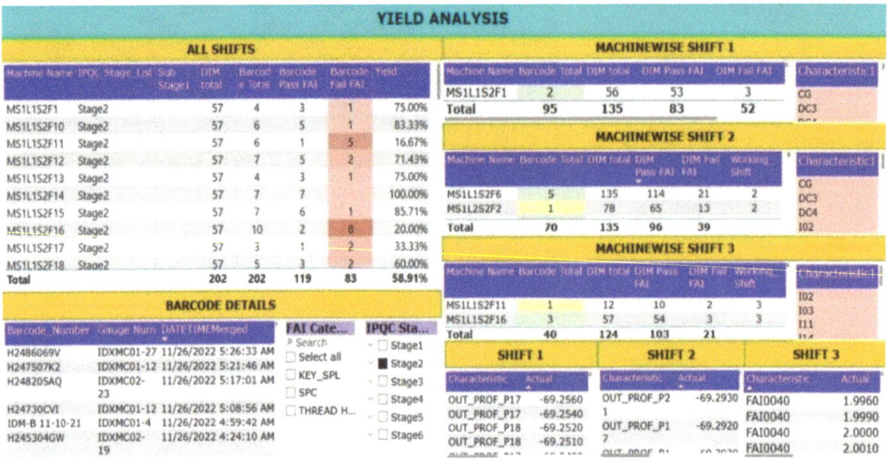

Fig. 3. Real-Time FAI Pass/Fail Classification with Barcode-Level Traceability and Yield Insights

4.3 Characteristic Heat Map

The Characteristic Heat Map enables fast detection of dimensional instability through process capability indices — CP and CPK. It presents a matrix mapping various characteristics against manufacturing stations or parts (e.g., Station 1, Ø10 $\pm$ 0.05), with color-coded risk indicators. Red cells indicate critical process issues (CP/CPK $<$ 1.00), yellow flags borderline capability (~1.10) and green cells represent statistically capable processes (CP/CPK $\geq$ 1.33). This visualization allows engineers to prioritize corrective actions by pinpointing low-performing dimensions or machines, guiding efforts in tooling adjustments, calibration or process re-engineering to achieve consistent tolerance compliance (Fig. 4).

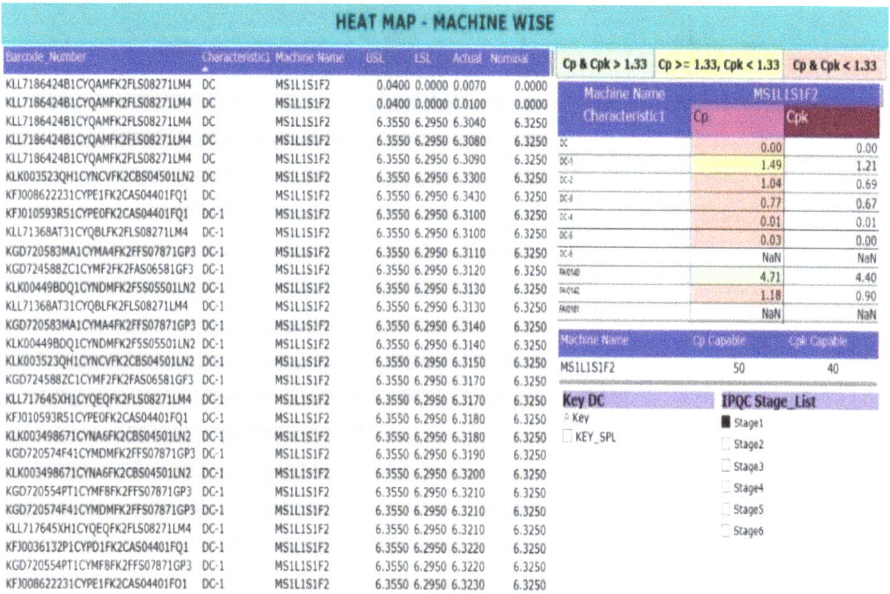

Fig. 4. Heatmap of Cpk and Cp Variations for Machine DC-1 Highlighting Process Stability Over Time

4.4 Failed FAI Analysis

The failed FAI Analysis module focuses on identifying patterns of repeated or compounded dimensional failures across different machines and part numbers. It displays Machine IDs, Part Numbers and corresponding Failure Counts to highlight top contributors to quality defects. For example, if Part 2304 shows frequent failures on Machine_05, it flags a high-risk zone. A color-coded visualization, with red bars indicating high failure frequency, supports quick prioritization of problem areas. This tool allows users to perform root cause analysis by drilling down into repeated partlevel and machine-level abnormalities, enabling proactive maintenance and corrective process interventions (Fig. 5).

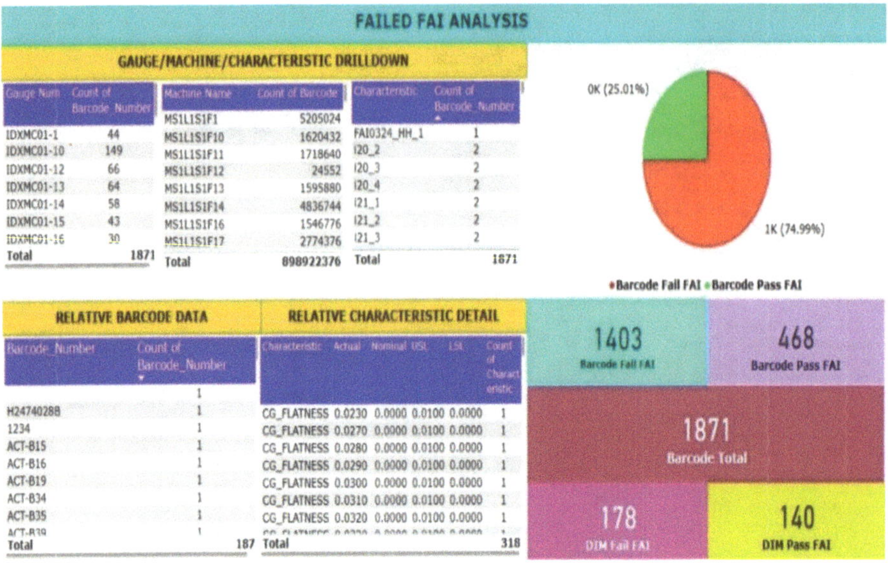

Fig. 5. FAI Failure Summary by Component with Frequency-Based Root Cause View.

4.5 QDIM Heartbeat

The QDIM Heartbeat dashboard serves as a live monitoring system for the connectivity and operational status of the Data Acquisition (DAQ) engine, CNC machines and quality systems. It provides indicators for DAQ Engine Running, QDIM Active, and CNC Machine Connected, along with real-time heartbeat timestamps. Any drop in communication or system fault is highlighted using a red dot or Not Active label, ensuring quick identification of connectivity issues. Machines like CNC_001 and CNC_003 are individually tracked for last sync activity, making this dashboard critical for maintaining uninterrupted data flow and ensuring continuous quality data capture from the shop floor (Fig. 6).

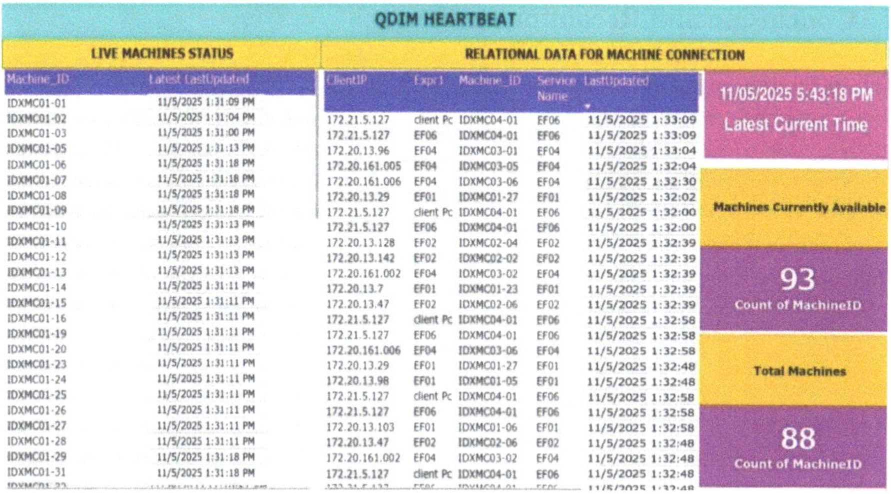

Fig. 6. Live System Health View Showing Machine Connectivity and Data Flow Status.

5 Observations and Results

The implementation of integrated analytics dashboards significantly improved quality decision-making, leading to a 15–20% increase in FAI yield. IPQC teams used realtime trend visualizations for shift planning, and machines with CpK below 1.33 were promptly flagged for maintenance. A key outcome was the identification of repeated gauge failures at a specific station, uncovered through Cpk heatmap trends demonstrating how visual analytics enabled early issue detection and process optimization. Additionally, operators and engineers aligned more effectively using shared dashboards, reducing manual reporting efforts and driving faster resolution cycles. The overall shift from reactive to proactive quality control enhanced production consistency and reduced rework.

6 Benefits to the Manufacturing Industry

Production Teams gain real-time visibility into machine performance, CpK trends, and failure hotspots, helping them optimize workflows and reduce unplanned downtime. Maintenance Teams use threshold-based alerts (like CpK < 1.33) to prioritize preventive actions, reducing breakdown frequency and improving asset reliability. Planning and Operations can leverage historical yield and trend data for better resource allocation, shift planning and scheduling of high-risk parts.

Management and Leadership obtain consolidated insights for KPI tracking, cost of quality analysis and strategic decision-making, leading to measurable ROI improvements.

Supplier Quality and Incoming Inspection can align their processes with plant-level data, ensuring early detection of supply-related deviations.

So, while quality teams directly benefit through faster RCA and better visibility, the ripple effect improves efficiency, communication and decision-making across the entire manufacturing ecosystem.

7 Conclusion and Recommendations

The integration of real-time analytics, statistical process control (SPC) indicators and visual dashboards into the quality assurance framework has proven to be a transformative strategy. The study underscores how data-driven visualization enhances process transparency, expedites decision-making and contributes to yield improvements. It is recommended that manufacturing enterprises adopt such integrated systems to minimize quality silos, reduce defect rates and enable proactive interventions. Future work should focus on embedding predictive AI models and realtime feedback mechanisms into the architecture to further enhance decision support and enable autonomous quality correction.

References

1. Patange, A., et al.: Integrated Quality 4.0 framework for quality improvement based on Six Sigma and machine learning techniques towards zero-defect manufacturing. Comp. Indus. Eng. **181**, 109094 (2023). https://doi.org/10.1016/j.cie.2023.109094
2. Banerjee, A., Pal, S., Mitra, S.: An IIoT machine model for achieving consistency in product quality in manufacturing plants (2021). arXiv preprint arXiv:2109.12964. https://arxiv.org/abs/2109.12964
3. Diallo, C., Rodriguez, C., Vallespir, B., Fortineau, V.: Towards a more efficient use of process and product traceability data for continuous improvement of industrial performances (2018). arXiv preprint arXiv:1810.13141. https://arxiv.org/abs/1810.13141
4. Saad, M., Ameer, M., Zaki, M.: Analysis of data mining process for improvement of production quality in industrial sector (2021). arXiv preprint arXiv:2108.07615. https://arxiv.org/abs/2108.07615

Personalized Education

Psychological and Pedagogical Principles of Medical Professionals Training in the Conditions of Transformational Changes Using Digital Technologies

Sofiia Dembitska[1(✉)], Iryna Kobylianska[1], Mariia Shostatska[2], Roman Salii[3], Oleksandr Kobylianskyi[1], Bogdan Pinaiev[1], and Olga Pinaieva[3]

[1] Vinnytsia National Technical University, 95 Khmelnitskoe Highway, Vinnytsia 21027, Ukraine
sofiyadem13@gmail.com

[2] Vinnytsia Academician D.K. Zabolotnyi Professional Medical College, Pirogova Street, 57, Vinnytsia 21000, Ukraine

[3] Vinnytsia Mykhailo Kotsiubynskyi State Pedagogical University, 32, Ostrozkogo Street, Vinnytsia 21000, Ukraine

Abstract. The article is devoted to the psychological and pedagogical principles of medical professionals training in the conditions of transformational changes using digital technologies. The purpose of the study is to determine the conceptual principles of the efficient integration of digital tools into medical education to increase the level of clinical, analytical and communicative competencies formation, as well as to develop recommendations for the optimal use of these technologies in the educational process. Analysis of modern research, regarding the application of digital technologies in medical education, in particular, main trends, achievements and problems was conducted. The study also reveals psychological and pedagogical mechanisms for the effective implementation of digital tools, contributing to the harmonious combination of practical training and the development of students' personal qualities. It was established that digital technologies allow adapting the educational process to the needs of each student, ensure safe and effective development of practical skills, stimulate critical thinking, responsibility and professional self-improvement. The integration of digital technologies into medical education is efficient tool for transforming the educational environment and preparing future medical professionals for the challenges of the modern healthcare system.

Keywords: medical education · digital technologies · distance learning · innovative educational technologies · improvement of educational process

M. E. Auer et al. (Eds.): IMCL 2025, LNNS 1954, pp. 225–234, 2026.
https://doi.org/10.1007/978-3-032-24998-2_19

1 Problem Statement

The current stage of society's development is characterized by large-scale transformational changes that cover all spheres of life, including the medical branch. The growing role of digital technologies, introduction of innovative educational environments and the need to quickly respond to new healthcare challenges require high-quality training of medical professionals who are able to effectively apply modern tools in their professional activities.

In this context, the psychological and pedagogical principles of training future medical professionals are of great importance. They determine not only the level of professional competence, but also the formation of critical thinking, emotional stability, the ability to work in a team and effectively interact with patients and colleagues in a digital environment. The use of innovative teaching aids based on information and communication technologies creates new opportunities for individualizing the educational process, increasing its interactivity and efficiency.

Therefore, the study of the psychological and pedagogical principles of medical workers training in the context of transformational changes using digital technologies is relevant and aimed at finding optimal ways to combine professional and personal training of future specialists with the needs of the modern healthcare system.

2 Analysis of Recent Research and Publications

Peculiaritie medical workers trainingv in the context of transformational changes in society are actively discussed in scientific discourse. Modern research emphasizes that the effectiveness of professional education of doctors largely depends on the integration of traditional pedagogical approaches with the latest digital technologies.

The works [1, 2] emphasize the need to combine cognitive, emotional-conative and value components in the formation of a professional personality. This is of particular importance for medical professionals, since their activities are associated not only with a high level of knowledge and practical skills, but also with the ability to empathize, support the patient and maintain psychological stability in stressful situations.

Modern health care system is undergoing significant changes, caused by both globalization processes and internal reforms. Many scientists believe [3, 4] that these changes make higher education institutions search for new models of professional training, focused on flexibility, interdisciplinarity and mobility. For medical education, this means not only updating the content of training, but also transformation of teaching methods and forms of educational process organization.

The widespread introduction of digital technologies into the educational process is considered by the scientists as a key condition for improving the quality of education. Thus, in the works [5–7], attention is focused on the possibilities of using electronic educational platforms, simulation technologies, virtual and augmented reality, distance courses and interactive simulators in the training of medical profecionals. They allow to bring the educational process closer to the conditions of real professional activity, form clinical thinking and practical skills in students. Thus, the analysis of scientific literature shows that the psychological and pedagogical principles of medical workers training should be considered in close connection with digital technologies, as they not only expand the possibilities of the educational process, but also form new requirements for the personality of the future specialist.

3 Theoretical and Methodological Principles of Medical Professionals Training Using Digital Technologies

The introduction of digital technologies into the educational process of higher education medical institutions is extremely relevant in modern educational context. This thesis is justified by a number of arguments:

1. Transformation of medical practice. Rapid development of digital technologies in medical field, in particular medical analytics, virtual and augmented reality, determines the need for specialists with a high level of digital competence. The integration of digital technologies into the educational program allows training medical specialists who are able to effectively use these tools in professional activities.
2. Adaptation to the needs of the student audience. Modern students belong to the "digital generation" (digital natives), which is organically integrated into the digital communications environment. The use of appropriate technologies in the educational process contributes to increasing their involvement, providing access to interactive and relevant educational materials and methods.
3. Improving the quality of educational services. Digital technologies are a tool for improving the quality of education in higher medical education institutions. They enable the use of various didactic approaches, such as virtual simulations, electronic educational resources, video lectures and interactive training modules, which effectively complement and expand the capabilities of traditional forms of teaching.
4. Compliance with labor market requirements. Modern labor market in medical branch puts forward increased demands on the digital literacy of the specialists. The integration of digital technologies into the educational process contributes to the formation of necessary skills in graduates, which increases their competitiveness and adaptability to changing conditions of the professional environment [13].

In this direction, VNTU has developed and implemented a set of measures, aimed at deepening scientific research. The results of these initiatives are highlighted in a number of publications [8–12]. In particular, for illustration, Fig. 1 presents the example of a student project, dedicated to the study of blood circulation processes. Figure 2 visualizes quantitative assessment of the ratio of blood circulation in the patient's affected arm compared to the healthy arm.

This project is an example of the application of thermography for medical diagnostics. It allows students to develop skills in integrating visual data with quantitative analysis, which is important for the objective assessment of physiological processes. The project has practical significance for the diagnosis of the diseases, associated with circulatory disorders. Participation in such projects gives students significant advantages that go beyond the usual academic program. They get the opportunity to use theoretical knowledge, gained at lectures to solve real scientific and engineering problems. Work on the project, from the data collection to their analysis and visualization, develops critical

Fig. 1. Implementation of a scientific project on thermographic studies of blood circulation by medical students

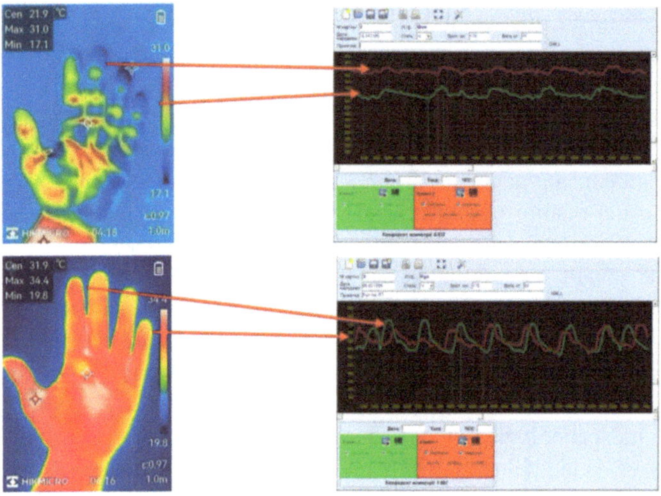

Fig. 2. Visualization of thermography results for assessing circulatory perfusion of the affected (top image) and healthy (bottom image) hands of patients

thinking, the ability to work with scientific sources and formulate conclusions. Such tasks simulate real projects in research institutes and industrial companies. This gives students a unique experience that makes them more competitive in the job market. Completed projects, the results of which are published or presented at conferences, become an important element of the student's professional portfolio, which can be decisive for the employment or admission to graduate school.

Students gain experience working with digital tools that are an integral part of modern medicine and science. This includes data analysis, the use of specialized software, and the visualization of complex medical processes. These skills are keys to their future careers, as they allow them to quickly adapt to new technologies and work effectively in a digital environment. Digital education prepares students to solve real medical problems in the digital age. They learn not only to accumulate information, but also to analyze it, find relationships, and apply it for diagnosis and treatment. This makes them more competitive and ready for innovations that are constantly emerging in the field of healthcare.

Analysis of scientific publications allows us to highlight the features of the digitalization of medical education, which are summarized in Table 1.

Table 1. Features of digitalization of medical education.

Aspect	*Areas of implementation*	*Expected result*
Accessibility of knowledge	Use of electronic educational platforms, distance courses, open educational resources	Expanding opportunities for learning regardless of place of residence and time
Practical training	Use of simulators, VR/AR, interactive simulators	Practicing clinical skills in safe conditions
Interactivity	Online seminars, virtual laboratories, digital case methods	Activation of cognitive activity, development of clinical thinking
Personalization of learning	Individual trajectories, adaptive learning systems	Taking into account the needs and level of each student preparedness
Communication and collaboration	Use of digital tools for group projects, telemedicine, forums	Development of teamwork, interpersonal and professional communications
Evaluation of results	Electronic testing, automated feedback systems	Objectivity, efficiency and transparency of assessment
Psychological aspect	The need to form a digital competence, resistance to information overload	Development of self-regulation, critical thinking and emotional stability
Continuity of education	Access to online databases, resources for postgraduate education	Support for the concept of lifelong learning for doctors

It is worth noting that medical education has its own specific features, which make it impossible, at least for now, to completely switch to digital technologies with the abandonment of traditional teaching methods. Despite all the advantages, distance learning cannot be the main form of education in higher medical education institutions. This is due to the fact that medical training involves the formation and improvement of practical, manual and clinical skills. These competencies cannot be fully acquired and developed without direct interaction with patients, work in real clinical conditions, as well as the performance of practical manipulations. Traditional teaching methods, which include clinical practice and simulation training, are indispensable for achieving the required level of professional training.

Thus, the optimal approach is the integration of digital technologies as a complement to traditional forms of education. This hybrid approach allows combining the advantages of both formats:

1. Digital technologies can be used for theoretical training, access to electronic libraries, virtual simulations that simulate certain clinical situations, as well as for practicing action algorithms.
2. Traditional methods are critically important for practical training: internships in medical institutions, work under the guidance of experienced clinicians, development of skills for communicating with patients and making decisions in real time.

The implementation of the hybrid model provides high-quality training of specialists who possess both modern digital literacy and the necessary practical skills that meet the requirements of modern medicine. Analysis of the theoretical and methodological foundations of the training of medical professionals shows that modern medical education should be based on the integration of traditional psychological and pedagogical approaches with the capabilities of digital technologies. Humanistic, competency-based, personality-oriented and activity-based approaches determine the basis for the formation of a professional personality of a physician, while digital tools provide expanded access to knowledge, modeling of professional situations and individualization of the educational process.

Transformational changes in society put forward new requirements for the training of physicians – the development of critical thinking, the ability to continuous learning, digital literacy and psychological resilience. In this context, the use of electronic educational platforms, simulation technologies and interactive teaching methods is not only an innovative tool, but also an important condition for improving the quality of professional training. Thus, the digitalization of medical education is not an additional element, but a system-forming factor that ensures the compliance of the training of future medical workers with modern healthcare needs.

4 Psychological Aspects of Medical Professionals Training, Using Digital Technologies

Modern educational paradigm requires comprehensive psychological support for all participants in the educational process. Digitalization, as one of the key drivers of change, significantly affects the psychological adaptation of the educational community, which can manifest itself in two main forms:

1. Psychological loyal-passive position. This approach is characterized by conservatism and evolutionary perception of innovations, where the educational community slowly integrates digital technologies, preserving traditional methods.
2. Psychological transformational-active position. This model involves radical changes in the educational environment due to the active implementation of digital tools and innovations.

Analysis of scientific publications [14–16] and generalization of own professional experience allows us to identify the following key psychological problems and tasks related to digitalization in medical institutions:

1. Enhancement of the psychological readiness of teachers to master digital competencies and effectively implement digitalization at all stages of training. This involves not only overcoming resistance to changes, but also actively forms a positive attitude towards the use of modern information and communication technologies in professional activities, developing motivation for constant self-improvement and the ability to adapt to dynamic transformations of the educational environment. The important aspect is also ensuring the emotional stability of teachers and supporting professional self-determination in the context of the digital transformation of learning.
2. Taking into account the psychological characteristics of the medical audience when designing and developing new digital educational technologies. It is necessary to take into account the high cognitive load and emotional stress, characteristic for medical education, as well as the peculiarities of students' perception, motivation and stress resistance. This involves the development of adaptive learning tools that contribute to the effective assimilation of the material, support psychological comfort and stimulate active involvement in the educational process in the digital environment.
3. Active implementation of virtual educational spaces that allow practicing key clinical and professional competencies. Such technologies include virtual surgical simulators, diagnostic simulators and interactive patient models, which provide a safe, controlled and adaptive environment for practical training. The use of such platforms allows to improve the quality of professional training, reduce the stressful component of the educational process and stimulate the development of critical thinking and decision-making in complex clinical situations.
4. Design of distance learning courses, based on the principles of productive distance learning. An important aspect is the integration of interactive educational tools, case methods, virtual simulations and tasks that simulate real professional situations. This approach ensures deep learning of the material, the development of critical thinking, the ability to make informed decisions and increases the readiness of future specialists for independent professional activity in the conditions of modern medical practice.
5. Intellectual and personal development. Intellectual and personal development of future medical specialists involves combining training in a virtual educational space with the development of their cognitive abilities and the formation of professional subjectivity. It is important that digital tools do not replace the process of thinking and decision-making, but act as a means that enhances critical thinking, responsibility and independence in professional activity. Such an approach contributes to a harmonious combination of intellectual growth, personal maturity and practical skills necessary for effective activity in modern medical practice.
6. Disclosure of psychological and pedagogical possibilities of infocommunication, multimedia and interactive technologies for imitating real clinical scenarios in the educational process. Use of such technologies allows to model complex professional situations, ensure safe practice of practical skills, stimulate critical thinking and making informed decisions, as well as increase the motivation and activity of students in mastering the educational material.
7. Identifying the potential of using artificial intelligence (AI) or its elements (e.g., chatbots for teaching clinical interviewing, intelligent systems for analyzing laboratory results) in distance learning. The use of AI contributes to the personalization of learning, operational feedback, automated assessment of knowledge, and the development

of analytical thinking of students. At the same time, it is important to ensure the integration of AI into training courses in such a way that technologies complement the pedagogical activities of the teacher, rather than replacing it, while maintaining the educational value and professional formation of future medical specialists.

8. Study of the mechanisms of group dynamics management in conditions of distance learning courses and its impact on the effectiveness of learning. This is especially important for the formation of teamwork skills in the medical field, where effective interaction and coordination of actions determine the quality of medical care. Analysis of group dynamics allows to optimize the organization of the educational process, increase student involvement, promote the development of communicative and leadership competencies, and also ensure effective teamwork in conditions of a distance educational environment.

Considering modern transformation processes caused by the digitalization of education, there is a need to clearly distinguish psychological aspects of its impact both on the individuality of the participant in the educational process and on the functioning of the educational community as a whole. To ensure the effective integration of digital innovations into medical education, it is important to take into account and purposefully apply a number of psychological guidelines (Table 2).

Table 2. Psychological guidelines for the digitalization of medical education.

Psychological guideline	*Content/characteristic*	*Expected effect*
Adaptive	Promotes rapid adaptation to new digital conditions	Reduction of psychological discomfort and stress
Motivational	Supports internal readiness and interest in digital learning	Increase of the level of students' involvement and activity
Regulatory	Helps control own psycho-emotional state in the digital environment	Reduction of the risk of digital stress and overload
Stabilizing	Provides psychological stability during a period of change	Formation of emotional balance and confidence
Self-organizing	Develops time and resource management skills	Optimization of educational activities, increasing productivity

The current vector of scientific research in the field of psychological and pedagogical aspects of the digitalization of education is the development of principles for the formation of psychological stability of students. This system should be based on the synergistic principles that involve the interaction and mutual reinforcement of all components to achieve a single goal – the formation of psychological readiness for innovative transformations.

The particular significance of this approach lies in the context of professional training of medical professionals. The implementation of telemedicine, digital simulations and electronic medical records requires not only technical skills, but also deep psychological readiness to work in conditions of constant technological change. Therefore, the development and implementation of effective psychological support programs is critically important for preserving the mental health of medical students and teachers, as well as for ensuring high quality medical education.

5 Conclusions

Training of medical professionals in the conditions of transformational change requires the integration of psychological and pedagogical approaches with modern digital technologies. The study shows that the effective use of virtual simulators, interactive trainers, distance courses and artificial intelligence tools contributes to the formation of key clinical, analytical and communicative competencies. Taking into account the psychological characteristics of the medical audience, management of cognitive and emotional load, stimulation of critical thinking and professional subjectivity of students increases the effectiveness of the educational process. Thus, the combination of traditional pedagogical principles with digital technologies ensures the comprehensive development of intellectual, personal and professional qualities of future medical specialists, which requires improving the quality of medical education in the conditions of modern transformational changes.

References

1. Bilyk, V.: Digitalization of the Educational Process as a Means of Increasing the Quality of Health-preserving Competence of Future Teachers in the Contest of Pandemic COVID-19. Pedagogy and Psychology of postmodernism: Values, Competence, Digitalization: Collective Monograph According to the scientific. In: Professor Anna Tsvietkova (Ed.). Aerzen : Heilberg IT Solutions UG (haftungsbeschränkt) InterGING Verlag., pp. 435–447 (2021)
2. Andriushchenko, K., Khaletska, A., Ushenko, N., et al.: Education process digitalization and its impact on human capital of an enterprise. J. Manage. Info. Decis. Sci. **24**(5), 1–9 (2021)
3. Seneviratne, T., Kodikara, K., Abeykoon, I., et al.: Perception of medical undergraduates on artificial intelligence in medical education: qualitative exploration JMIR. Med. Educ. **11**, e73798 (2025). https://doi.org/10.2196/73798
4. Jaleel, A., Aziz, U., Farid, G., et al.: Evaluating the Potential and Accuracy of ChatGPT-3.5 and 4.0 in Medical Licensing and In-Training Examinations: Systematic Review and Meta-Analysis JMIR Med Educ **11**, e68070 (2025). https://doi.org/10.2196/68070
5. Bouhoula, M., Kacem, I., Ghenim, A., et al.: The room of errors as an innovative simulated environment to enhance the management of blood exposure accidents: a tunisian pre-experimental study. Adv. Med. Educ. Pract. **16**, 1641–1650 (2025). https://doi.org/10.2147/AMEP.S530487
6. Scaglione, M., Calandrino, A., Puzone, L., et al.: Enhancing pediatric residency training through peer-education based gamified simulation. Adv. Med. Educ. Pract. **16**, 1621–1626 (2025). https://doi.org/10.2147/AMEP.S518814
7. Al-Kubaisi, F.E., Dargham, S.R., Manyama, M.F.: Enhancing neuroanatomy education for medical students through the development of online modules. Adv. Med. Educ. Pract. **16**, 1601–1608 (2025). https://doi.org/10.2147/AMEP.S541772

8. Pavlov, V.S., Zabolotna, N.I., Polishchuk, V.S., et al.: Biometric identification of photoplethysmographic signals in the clinic of maxillofacial surgery using deep learning technologies. Scientific Works of Vinnytsia National Technical University **1**, 12 (2025). https://doi.org/10.31649/2307-5376-2025-1-109-120
9. Korenivska, O., Benedytskyi, V., Nikitchuk, T., et al.: Application of optical methods for measuring physiological parameters in the construction of telemedicine systems for the diagnosis of infants and children. Proceedings Volume Photonics Applications in Astronomy, Communications, Industry, and High Energy Physics Experiments **2024**, 134000I (2024). https://doi.org/10.1117/12.3058675
10. Tymchenko, L., Kokriatska, N., Tverdomed, V., et al.: Parallel-hierarchical optical network as a model of natural neural network. Proc. SPIE 12985, Optical Fibers and Their Applications 2023, 129850D (2023). https://doi.org/10.1117/12.3023432
11. Ushenko, O, Pavlyukovich, N., Khukhlina, O., et al.: Mueller matrix and laser induced imaging of the myocardium histological sections in the diagnosis of long-term consequences of COVID19. Proceedings Volume Photonics Applications in Astronomy, Communications, Industry, and High Energy Physics Experiments **2024**, 1340004 (2024). https://doi.org/10.1117/12.3054800
12. Khriienko, K.S., Chyzh, I.G., Kolobrodov, V.G., et al.: Computer-integrated method for determining the acceptable errors in measuring aphakia eye parameters in the treatment of cataracts. Proc. SPIE 12040, Photonics Applications in Astronomy, Communications, Industry, and High Energy Physics Experiments 2021, 120400A (2021). https://doi.org/10.1117/12.2607824
13. Zhai, X., Nie, Y.: To Evaluate Interdisciplinary Integration in Medical Education Through a Pulmonary Fibrosis Case Study. Adv. Med. Educ. Pract. **16**, 1529–1536 (2025). https://doi.org/10.2147/AMEP.S529501
14. Sverdrup, O., Jensen, T., Solheim, S., et al.: Training auscultatory skills: computer simulated heart sounds or additional bedside training? A randomized trial on third-year medical students. BMC Med. Educ. **10**, 3 (2010). https://doi.org/10.1186/1472-6920-10-3
15. Azad, R., Shakour, M., Moharami, N.: Designing an evaluation tool for evaluating training programs of medical students in clinical skill training center from consumers' perspective. BMC Med. Educ. **24**, 502 (2024). https://doi.org/10.1186/s12909-024-05454-7
16. Si, J.: Using cognitive load theory to tailor clinical reasoning training for preclinical medical students. BMC Med. Educ. **24**, 1185 (2024). https://doi.org/10.1186/s12909-024-06111-9

A Proposed Modular Instructional Design Framework for Blended Learning Approach

Archi Choudhary(✉) and Rajeev Chatterjee

National Institute of Technical Teachers' Training and Research, Kolkata, India
archichoudhary254@gmail.com

Abstract. Blended Learning (BL) has become a widely adopted educational model, merging traditional face-to-face instruction with online learning tools and techniques. It offers many advantages, including flexibility and structured interaction. During the pandemic and post-pandemic era, there is a requirement to deliver robust, adaptable, and context-sensitive instructional design frameworks that will lead to a paradigm shift in the Teaching and Learning System (TLS). Despite the availability of established models like Analysis, Design, Development, Implementation, and Evaluation (ADDIE) and conceptual frameworks like Technological Pedagogical Content Knowledge (TPACK), there may be substantial research gaps available to fully support the dynamic demands of BL approaches. This research proposes a comprehensive instructional design framework integrating the existing available standards and models for enhanced learning experience, promoting motivation, engagement, and reflection among learners. The framework envisions a modular digital platform where TPACK informs the design phase, and Self-Regulated Learning (SRL) strategies are embedded throughout development, implementation, and evaluation. This proposed model, currently in the early stage of development, aims to empower instructors to design more coherent, adaptive, and learner- centered BL experiences.

Keywords: Technology Enabled Learning (TEL) · Teaching and Learning System (TLS) · Blended Learning (BL)

1 Introduction

Blended Learning (BL) has emerged as a transformative and adaptable approach within modern Teaching-Learning systems (TLS), This responds to the growing need for flexible, personalized and technology integrated education. It offers a balanced combination of traditional face-to-face instruction with Technology Enhanced Learning (TEL) experiences, allowing synchronous interactions and asynchronous engagements. This flexibility allows instructors to adapt to diverse learning styles, updated content requirements, and learning outcomes. While these models offer pliability and engagement, many BL implementations fall short due to the absence of a coherent instructional strategy. This is due to the fact that the existing instructional design approaches tend to emphasize either on the technological delivery mechanisms or the content itself, often neglecting

M. E. Auer et al. (Eds.): IMCL 2025, LNNS 1954, pp. 235–249, 2026.
https://doi.org/10.1007/978-3-032-24998-2_20

the pedagogical coherence and learner action essential to meaningful learning. As a result, instructors may struggle to develop cohesive learning experiences that integrate content, pedagogy, and technology while fostering SRL behaviors in students/learners.

Instructional design is the systematic process of creating effective, engaging, and goal-oriented learning experiences. It plays a critical role in aligning content, pedagogy, and technology to meet specific learning outcomes. By incorporating structured models like ADDIE, content-technology alignment through TPACK, and learner autonomy via SRL, a modular instructional design framework can empower instructors to design meaningful, flexible, and learner-centered BL environments. This integrated approach addresses both the complexity and the potential of modern education.

The three potential elements of Instructional System Design (ISD) have been discussed in the following paragraph:

- ADDIE

It provides a systematic process for instructional planning and course development. Each phase supports logical progression, from understanding learner needs to evaluating learning outcomes. Figure 1 illustrates the iterative structure of ADDIE.

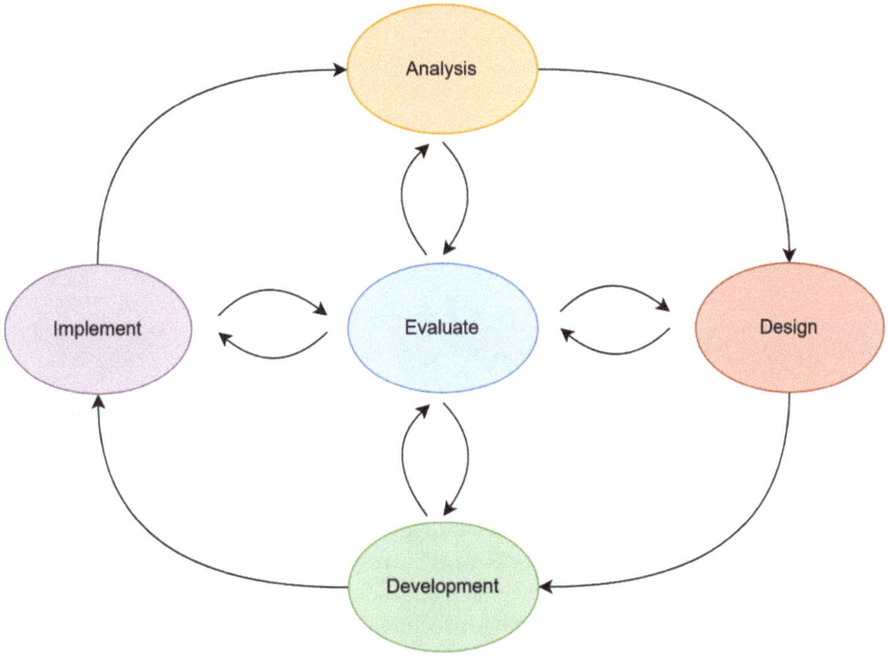

Fig. 1. ADDIE Model

- TPACK

It emphasizes the interplay of technology, pedagogy, and content. It ensures that instructional decisions are not just tool-driven but pedagogically sound and content-aligned. Figure 2. Illustrates the TPACK Model.

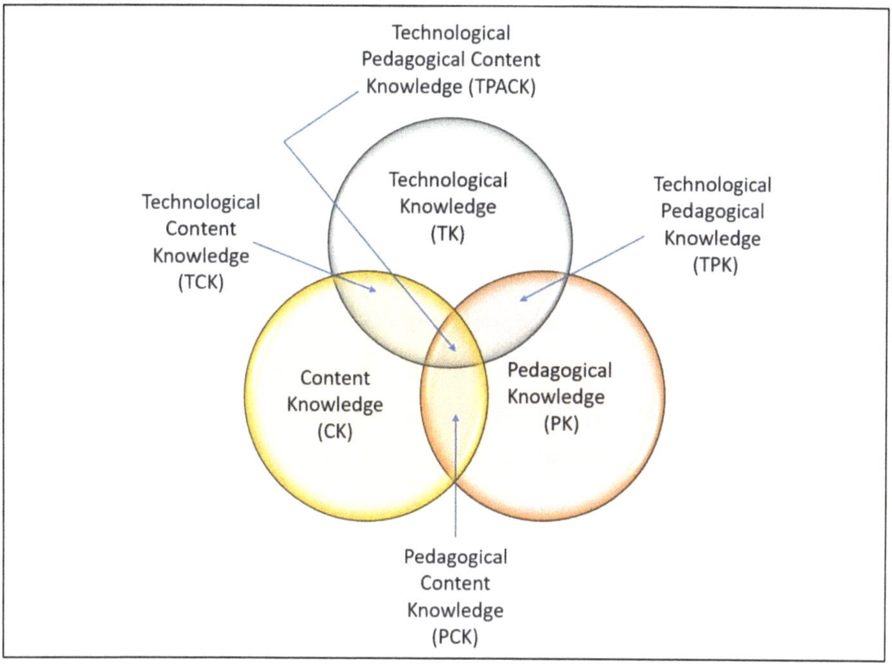

Fig. 2. TPACK Model

- SRL

It introduces the learner's ability to set goals, monitor progress, and reflect on learning. SRL is particularly important in blended learning environments where students are expected to manage parts of their learning independently.

By synthesizing these models into a unified modular framework, the instructional design becomes more holistic, scalable, and reflective of contemporary pedagogical needs.

The paper is structured as follows, Section 2 contains the review of the existing work. The proposed framework has been laid down in Section 3. Results and Discussion are given in Section 4. The conclusion and future scope are given in Section 5.

2 Review of the Existing Work

In this section, the authors have provided a comprehensive view of the existing research activities that are been carried out in the recent past in the related domain.

Kumar et al. present an extensive analysis of blended learning tools and practices, classifying them across platforms, pedagogical strategies, and learner engagement techniques [1]. However, while they thoroughly map the tool ecosystem, they lack a structured instructional design model integrating pedagogical frameworks.

Kintu et al. explore how student characteristics and design features influence the effectiveness of blended learning [2]. They provide data-backed validation that design quality significantly impacts learning outcomes, but the study stops short of offering a concrete design framework that instructors can replicate or scale.

Liu et al. review major educational theories like constructivism and connectivism and their application in hybrid environments in their book chapter [3]. They did not provide a practical way to implement these theories in real-world learning design.

Alammary highlights the need for intentional selection of design elements [4]. It lacks a systematic approach to promote sustainability in student engagement and success, a modular process to guide such selection.

Papanikolaou et al study leverages learning design to build TPACK competency in blended teacher training [5]. It provides an excellent case for TPACK integration but is focused primarily on teacher professional development rather than broader course design.

Parra et al present a redesign of a teacher education course using technology-enhanced methods, emphasizing the importance of context in technology integration [6]. While it offers a strong case for change in teacher preparation, it lacks a structured design model that ensures consistent implementation across contexts.

Adri et al. apply the ADDIE model to design a project-based blended learning environment [7]. This uses a production-based approach. It demonstrates the adaptability of ADDIE in vocational learning but focuses primarily on content development and sequencing.

González applies the ADDIE model to teacher training, offering a structured approach [8]. However, its isolated use lacks pedagogical richness and ignores learner autonomy essential for modern blended learning.

Baliya et al. propose an integrated TPACK model for blended learning in the post-COVID era [9]. This shows how technology and pedagogy could be merged for effective remote teaching. While this was impactful, the model remains primarily conceptual, without concrete workflow tools or design modules.

Kavitha & Anitha, in their case study, applied a TPACK-based model to improve engineering education and evaluate it using the Kirkpatrick Evaluation Model [10]. It effectively combines content and pedagogy but centers more on evaluation outcomes rather than on the design process itself.

Waluyo & Nuraini propose an instructional model using TPACK [11]. The model is integrated with inquiry-based learning to improve problem-solving skills. The model is strong in combining active learning and technology and focuses on one pedagogical strategy and does not provide flexibility for various learning contexts.

Sarac et al. highlight the effectiveness of a blended learning-based professional development program grounded in understanding by Design to enhance preschool teachers' capacity to promote self-regulation in children [12]. However, the study is limited by the absence of long-term follow-up to evaluate sustained teacher practices and classroom impact.

Yang et al. introduce a Learning Analytics Dashboard (LAD)–based self-regulated learning (SRL) approach to improve university students' engagement with e-books in blended learning environments [13]. The approach significantly enhanced learners' academic outcomes, SRL awareness, self- efficacy, and e-book engagement compared to conventional e-book use.

Benson et al. conducted a mixed-methods study to assess a Learning Design Support Tool designed to improve LMS course quality and accessibility in higher education [14]. The results highlight the importance of clear course structure, active learning, and clear instructions. However, the study mainly reflects instructor and reviewer opinions, lacking student feedback on course usefulness and engagement.

2.1 Research Gap and Problem Definition

In this section, notable research gap related to literature review has been presented as follows Kumar et al. thoroughly analyzed blended learning tools and practices but then tools are seldom aligning to perspective of ISD along with pedagogical framework. Kintu et al presented how student characteristics and design features affect outcomes but do not present a concrete replicable framework for instructional planning [2].

In order to sustain engagement, Alammary emphasis the importance of thoughtful component selection [4]. However, the manuscript does not propose a structured modular implementation guide. Ballia et al proposes a TPACK based approach to blended learning that aligns pedagogy and technology. However, there is a lack of concrete workflow merely at a conceptual level without design modules that instructors can implement directly [9].

From the aforesaid discussion it has been revealed that there is a notable gap in structured instructional design framework that effectively integrate pedagogy technology and learner centered practices. Further these limitations highlight the need for a unified modular instructional design framework that not only addresses integration of technology with pedagogy but also support replicability scalability and learners' autonomy.

In order to address this research gap, there is a requirement for design and development of a modular instructional design framework, which will integrate a structured model: ADDIE, with pedagogical and technical alignment: TPACK providing learner autonomy strategies using SRL in a cohesive manner.

3 Proposed Framework

In this study, the research problem statement in Section 2.1 has been considered and a framework has been proposed as a solution. The proposed framework strategically integrates three foundational models to address the complex demands of instructional design in blended learning environments. ADDIE provides a rigorous, phase-driven structure that ensures instructional activities are systematically analyzed, designed, developed, implemented, and evaluated. Building on this foundation, TPACK ensures that instructional design decisions thoughtfully align pedagogy, technology, and subject matter, promoting coherence and contextual relevance in the learning experience. To further enhance learner engagement and effectiveness, principles of SRL are incorporated across each phase, emphasizing goal-setting, self-monitoring, and reflective practices. Collectively, this integrated approach fosters structured, technology-enhanced, and learner-centered educational experiences.

The ADDIE model, TPACK framework, and SRL prompts were purposefully integrated into the proposed framework to ensure a comprehensive and effective instructional design process tailored for blended learning environments.

- The ADDIE model was employed for its structured yet flexible approach, offering iterative evaluation at every phase—from analysis to evaluation—thereby enabling continuous refinement of the instructional process.
- The TPACK framework was integrated to guarantee the meaningful alignment of technology with pedagogical methods and subject content, ensuring that digital tools are selected not in isolation but as part of a coherent educational strategy.
- In addition, SRL prompts were embedded to actively encourage learners to plan, monitor, and reflect on their learning during asynchronous tasks, promoting autonomy, intrinsic motivation, and deeper engagement with the content. This integration supports a learner- centered approach while maintaining instructional coherence and adaptability.

3.1 Terminal Components and Pedagogical Alignment

According to the ADDIE Model of ISD, the proposed design follows five key phases, also known as five terminals, each of which according to our proposed design has been explained below:

i. **Analysis Terminal**

   The purpose of this terminal is to gather learner-specific and context-specific information to guide the instructional design process. The components of this terminal are as follow:

   - Learner Profile: Captures details like prior knowledge, learning styles.
   - Context Constraints: Identifies infrastructural or environmental limitations.
   - Learning Objectives: Clearly defines what the learner should achieve.
   - Instructional Theory: Suggests pedagogical approaches suited for the learner profile.

ii. **Design Terminal**

   This terminal plans the instructional content and structure using TPACK principles, including the following components:

   - Flowchart of Content: Outlines how content will be sequenced and delivered.
   - TPACK-Informed Tool Choices: Ensures the selection of technology aligns with pedagogy and content.
   - Assessment Strategies: Defines formative and summative assessment plans.
   - Visual Layout Plan: Provides design guidelines for digital interfaces and course presentation.

iii. **Development Terminal**

The purpose of this terminal is to create learning materials and integrate SRL features, the components of this terminal are as follow:

- Final Learning Modules: Actual development of content resources (videos, PDFs, simulations).
- SRL Prompts & Scaffolds: Embedded self-regulation supports like reflection questions. For Example,

  Goal-Setting Prompt:

  "What are two specific goals you want to achieve by the end of this module?"

  Monitoring Prompt:

  "Pause and rate your current understanding of this topic on a scale of 1–5."

  Reflection Prompt:

  "What was the most challenging concept in this lesson, and how did you attempt to overcome it?"

  Self-Evaluation Prompt:

  "Compare your performance on this activity with your initial goals. Did you achieve them?"

  Motivation/Strategy Prompt:

  "If you get stuck, list one alternative strategy you could try before seeking help."
- Assessment Activities: Quizzes, tasks, and interactive exercises that align with objectives.

iv. **Implementation Terminal**

This terminal delivers the course and monitors real-time learner activity; the components of this terminal are as follows:

- Learner Interaction Data: Tracks click, time spent, resource usage.
- Reflections & Self-Assessments: Learner-submitted self-evaluations.
- Task Submissions & Quiz Scores: Collected as evidence of progress.
- Instructor Observations: Notes on learner performance and engagement.

v. **Evaluation Terminal**

The Evaluation terminal analyzes outcomes and provides feedback for continuous improvement, including the following components:

- Performance Trends: Patterns in learner success or struggle areas.
- Reflection Insights: Qualitative data from learner reflections.
- Tool Usage Reports: Analytics on how learning tools were used.
- SRL Dashboard: Summarizes learner self-regulation activities and progress.

The illustration of the various components of each terminal, along with their directional flow, is shown in Figure 3. This has been created based on the ADDIE Instructional Design model.

Example Module: Introduction to Programming with Python

- Analysis: Grade 10 beginners; objective is to understand variables, input/output, and basic control flow; context to school LMS with coding sandbox.

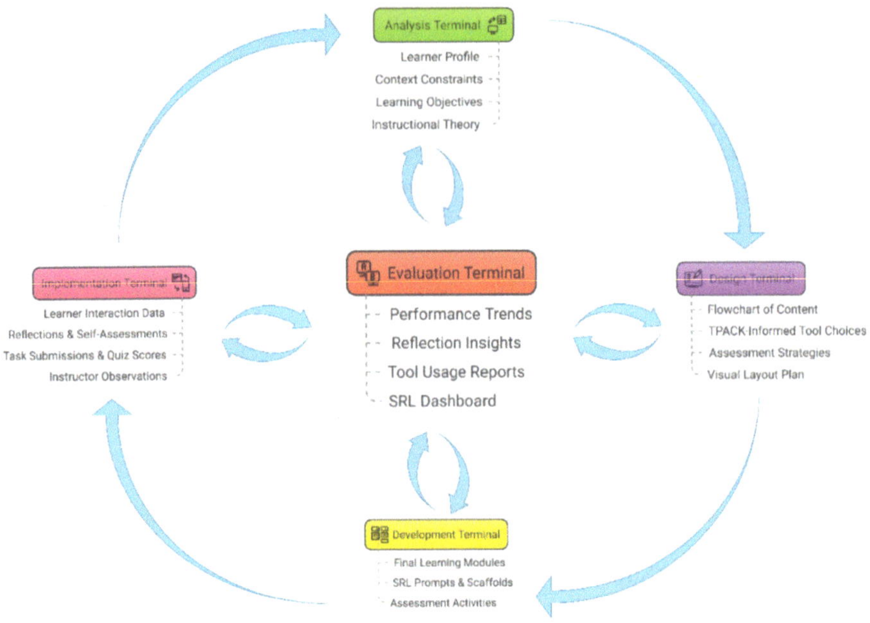

Fig. 3. Terminals and their various components

- Design (TPACK): Flipped classroom; pre-class video + coding examples; in-class debugging tasks; technology (sandbox), pedagogy (hands-on), content (Python basics).
- Development: Screencasts, guided coding tasks, and quizzes created; SRL prompts embedded (e.g., "What steps will you try if your code fails?").
- Implementation: Learners accessed materials via LMS, practiced in sandbox, submitted code, and reflected on strategies; performance data tracked automatically.
- Evaluation: Code correctness, quiz scores, and reflections analyzed; tool usage logs reviewed; SRL dashboard summarized learner progress.

Figure 4 presents a use-case that explains the proposed modular instructional design framework for BL. The system involves three key actors—Instructor, Learner, and the Learning Management System (LMS)—who play distinct but interconnected roles in ensuring the instructional design process is effective, personalized, and scalable.

The instructor is the primary user of the framework and plays a central role in orchestrating the entire instructional design process. Based on inputs from the analysis phase, the instructor identifies learner profiles, sets learning objectives, selects appropriate pedagogical strategies, and integrates suitable technological tools. During the design and development phases, the instructor creates learning materials, embeds SRL prompts, and prepares assessment strategies aligned with the course outcomes. As the learning experience unfolds, the instructor also monitors learner engagement and evaluates performance data to make iterative improvements, ensuring the learning journey remains meaningful and aligned with desired outcomes.

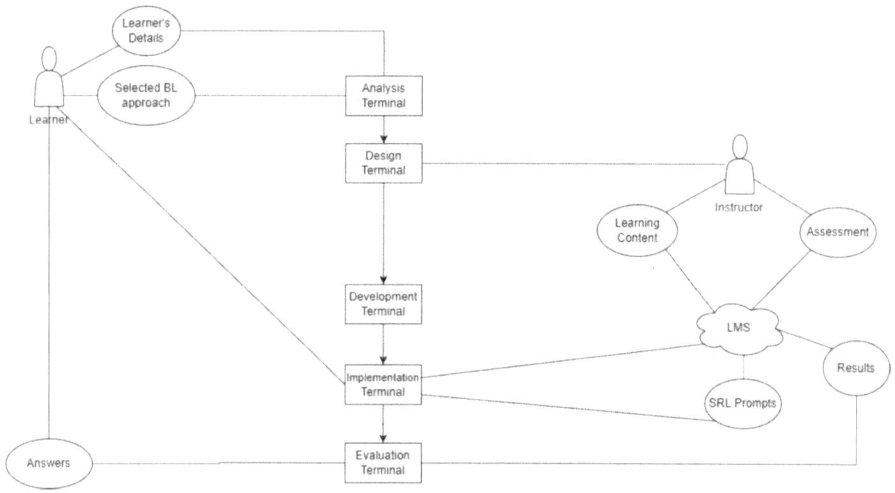

Fig. 4. Use-case diagram for the Proposed System

The Learner is the end-user who interacts with the learning modules and resources created within the framework. Learners are central to the blended learning approach, particularly due to its emphasis on learner autonomy and self-regulation. They engage with both synchronous and asynchronous components of the course, actively participate in learning activities, set personal learning goals, reflect on their understanding, and complete formative and summative assessments. Nevertheless, the learners were seldom involved in designing and configuring the framework, still the data related to behavioral aspects, feedback mechanism and based on their performance becomes critical input for enhancing instructional strategies and future iteration.

The LMS acts as an external but indispensable system used during the implementation and evaluation terminals of the framework. It serves as the delivery platform that hosts the learning content, enables interaction, and facilitates tracking of learner activities and progress. It also generates reports that support instructors in monitoring engagement levels and assessing learner outcomes. Although the LMS is not involved in the internal logic or modular structure of the framework itself, it is essential for deploying the learning experience and evaluating its impact in real educational environments.

The above three actors form a cohesive system in which content design, learner interaction, and data-driven refinement are deeply intertwined, enabling the framework to support adaptive, learner- centered blended learning environments. Section 3.2 presents an algorithmic approach for instructional design.

3.2 An Algorithmic Approach Towards the Proposed Instructional Design

Our instructional design for the Blended Learning approach consists of five terminals, each interconnected with the others via its components. Below is an algorithmic approach that explains how the instructional design works.

Algorithm1: Terminal to terminal workflow of the learning process.
Input: Prerequisite of succeeding Terminal
Output: Insights from preceding Terminal

```
Begin
{
  switch(action)
  {
    case 1: // Analysis Terminal
      // Process learner profile, context constraints, objectives, and instructional theory
      learner_profile = processLearnerProfile();
      context_constraints = processContextConstraints();
      learning_objectives = processLearningObjectives();
      instructional_theory = processInstructionalTheory();

      // Pass results to Design Terminal
      DesignTerminal(learner_profile, context_constraints, learning_objectives,
instructional_theory);

    case 2: // Design Terminal
      // Create a flowchart and align tools, assessment, and layout
      flowchart = createFlowchart();
      tool_compatibility = selectTPACKTools();
      assessment_strategies = alignAssessment();
      layout = designVisualLayout();

      // Pass results to Development Terminal
      DevelopmentTerminal(flowchart, tool_compatibility, assessment_strategies, layout);

    case 3: // Development Terminal
      // Build final learning modules, integrate SRL prompts, ensure mobile compatibility
      learning_modules = buildLearningModules();
      SRL_prompts = embedSRLPrompts();
      mobile_content = prepareMobileContent();
      assessment_activities = createAssessmentActivities();

      // Pass results to Implementation Terminal
      ImplementationTerminal(learning_modules, SRL_prompts, mobile_content,
assessment_activities);

    case 4: // Implementation Terminal
      // Deploy learning modules, track learner interaction, and assess activities
      deployInLMS();
      trackLearnerInteraction();
      deploySRLActivities();
      assessLearnerActivities();

      // Pass results to Evaluation Terminal
      EvaluationTerminal(learner_interaction_data, task_submissions, instructor_feedback);

    case 5: // Evaluation Terminal
      // Analyze learner performance, feedback, and tool usage
      performance_trends = analyzePerformance();
      reflection_insights = gatherReflectionInsights();
      tool_usage_reports = collectToolUsageReports();

      // Feedback loops back to Analysis Terminal for next iteration
      AnalysisTerminal(performance_trends,reflection_insights, tool_usage_reports);
  }
}
End
```

4 Results and Discussion

As the BL approaches, due to its learner-centered nature and flexibility, encompasses various models, each with its distinct properties and characteristics. The proposed modular framework provides a systematic, adaptable solution that seamlessly aligns learning objectives, pedagogical strategies, and technological tools within a unified, phase-driven workflow. By embedding flexibility as a foundational principle, the framework facilitates easy adaptation across diverse BL models, including rotation-based, flipped, lab-based, and enriched virtual formats. Table 1 illustrates the detailed workflow across the different terminals/phases, tailored for various BL models.

This modular, terminal-based approach enables scalability by supporting personalized pathways without overwhelming instructors with design complexity. It allows for dynamic modifications based on learner needs, contextual constraints, and model preferences. As a result, blended learning under this framework transcends fragmented, tool-centric practices and moves toward the creation of cohesive, reflective, and sustainable educational experiences that are responsive to both individual learner profiles and broader educational goals.

A comparative study is presented in Table 2, which systematically differentiates between the standalone TPACK and ADDIE models and the proposed integrated model, thereby elucidating their respective strengths, inherent limitations, and the enhanced pedagogical coherence, learner-centeredness, and technological alignment offered by the integrated framework.

Learner autonomy is also actively supported through the integration of SRL strategies, empowering them to set learning goals, monitor their progress, engage in metacognitive reflection, and take ownership of their educational journeys. The application of

Table 1. Workflow across the different terminals for various BL models.

Blended Learning Models	Phases				
	Analysis	Design	Development	Implementation	Evaluation
Rotation Model	Identify learners' readiness for rotating between stations/modalities	Design time schedules and learning paths for each station	Build modular, station-specific content	Deploy rotational timetables and station monitoring tools	Track learning outcomes per station; adjust rotation speed and structure
Flex Model	Assess learner independence and digital access levels	Design flexible pathways where students control pace and progression	Develop self-paced modules, with teacher interventions embedded	Deploy content in LMS with real-time feedback tools	Analyze time logs and learning mastery patterns; adjust content pacing

(continued)

Table 1. (*continued*)

Blended Learning Models	Phases				
	Analysis	Design	Development	Implementation	Evaluation
A La Carte Model	Mapping the content learners choose online vs. face-to-face	Customize independent learning contracts based on selected courses	Develop digital-only course units; prepare educator-support guidelines	Launch selected modules; integrate optional face-to-face tutoring sessions	Review course-specific achievements; survey learner satisfaction
Enriched Virtual Model	Analyze learner motivation for part-time virtual learning	Plan hybrid courses with strong virtual anchors and scheduled meetups	Develop rich online course content with occasional synchronous webinars	Schedule periodic face-to- face sessions while monitoring virtual attendance	Evaluate engagement metrics in both environments; adjust meetup frequency
Face-to-Face Driver Model	Assess how the traditional classroom can best leverage online supplements	Design lessons that embed online tools as homework or reinforcement	Build enhancement modules	Conduct face-to- face sessions; assign online tasks post-lesson	Evaluate blended homework effectiveness; adapt content delivery mix
Flipped Classroom Model	Profile learners' readiness for self-driven pre-class learning	Design pre-class activities and in-class active tasks	Create asynchronous learning resources + synchronous application tasks	Deliver pre-class material via LMS; run in-class active learning sessions	Track pre-class completion rates and in-class performance; refine activity balance
Lab Rotation Model	Identify subject areas best suited for digital lab time	Schedule rotations through physical labs and online labs	Develop interactive simulations or virtual lab experiments	Implement lab management tools for station tracking and progress logging	Evaluate lab skill acquisition through reports and quizzes; adjust lab station setup

the TPACK framework ensures that technology integration is pedagogically sound and content-aligned, avoiding the fragmentation often seen in blended learning implementations. Meanwhile, the ADDIE model ensures systematic instructional sequencing, guiding educators logically from learner analysis through evaluation.

Table 2. Workflow across the different terminals for various BL models.

Dimension	Standalone ADDIE	Standalone TPACK	Proposed Integrated Framework (ADDIE + TPACK + SRL)
Procedural Approach	Provides a systematic, phase-driven approach. Strong in sequencing but may become rigid or overly linear	Does not prescribe a process, relies on the educator's ability to integrate tech, pedagogy, and content	Combines ADDIE's structure with TPACK's knowledge alignment, ensuring both process discipline and contextual flexibility
Technology Integration	Technology often treated as an add-on, not embedded into each design phase	Core focus on technology–pedagogy–content balance, but lacks step-by-step workflow	Technology is embedded meaningfully at every stage via TPACK, avoiding tool-centric integration
Pedagogical Coherence	Can ensure objectives and content are aligned, but may underplay pedagogy in favor of content/sequence	Strong emphasis on pedagogy–content–technology interaction but less guidance on instructional sequencing	Aligns pedagogical intent with technology and content while maintaining structured sequencing
Learner Autonomy	Limited; learners are recipients of content with minimal self-regulation focus	Indirect; encourages student-centered approaches but does not explicitly address self-regulation	Explicit SRL prompts foster goal-setting, monitoring, and reflection, promoting learner autonomy and motivation

(*continued*)

Table 2. (*continued*)

Dimension	Standalone ADDIE	Standalone TPACK	Proposed Integrated Framework (ADDIE + TPACK + SRL)
Assessment & Feedback	Includes evaluation but often at the end; limited formative integration	Acknowledges assessment alignment but does not specify mechanisms	Embeds assessment strategies in Design & Development + SRL dashboard + continuous evaluation, supporting iterative improvement

5 Conclusion and Future Scope

The proposed modular instructional design framework offers a structured, flexible approach that promotes reflective design practices while supporting diverse learner profiles, technological contexts, and pedagogical requirements. By aligning ADDIE, TPACK, and SRL models within a terminal- based interface, the framework moves beyond fragmented practices to foster cohesive and sustainable educational experiences.

Future work will focus on empirical validation through classroom-based implementations and mobile learning scenarios. Planned studies will involve quantitative and qualitative assessments of the framework's impact on learner engagement, instructional efficiency, and overall learning outcomes. Additionally, iterative refinements based on user feedback from instructors and learners will be incorporated to enhance usability, adaptability, and scalability across different educational settings and disciplines.

Acknowledgment. This research activity was carried out in the Department of Computer Science and Engineering at NITTTR, Kolkata. The authors express their gratitude to the members of the faculty and staff of the Department of Computer Science & Engineering for their constant support and encouragement.

References

1. Kumar, A., Krishnamurthi, R., Bhatia, S., et al.: Blended Learning Tools and Practices: A Comprehensive Analysis. IEEE Access **9**, 85151–85197 (2021)
2. Kintu, M.J., Zhu, C., Kagambe, E.: Blended learning effectiveness: the relationship between student characteristics, design features and outcomes. Int. J. Educ. Technol. High Educ. 14 (2017)

3. Liu, M., Zhao, G., Zhong, Z., et al.: Theoretical Foundations for Blended Learning. Handbook of Educational Reform Through Blended Learning, 1–44 (2023)
4. Alammary, A.S.: Optimizing components selection in blended learning: toward sustainable students engagement and success. Sustainability **16**, 4923 (2024)
5. Papanikolaou, K., Makri, K., Roussos, P.: Learning design as a vehicle for developing TPACK in blended teacher training on technology enhanced learning. Int. J. Educ. Technol. High Educ. 14 (2017)
6. Parra, J., Raynor, C., Osanloo, A., Guillaume, R.O.: (Re)Imagining an Undergraduate Integrating Technology with Teaching Course. TechTrends **63**, 68–78 (2018)
7. Adri, M., Wahyuni, T.S., Ganefri, S.Z., Jama, J.: Using ADDIE Instructional Model to Design Blended Project-Based Learning based on Production Approach. Int. J. Adv. Sci. Technol. **29**(06), 1899–1909 (2020). Retrieved from http://sersc.org/journals/index.php/IJAST/article/view/12894
8. Morales González, B.: Instructional design according to the ADDIE model in initial teacher training. Ap **14**, 80–95 (2022)
9. Baliya, J., Thappa, S., Shikha, D.: Efficacy of TPACK Framework through Blended Learning Approach: An Innovative IntegratedModel in Post COVID-19 Era. **11**, 1–11 (2022)
10. Kavitha, D., Anitha, D.: TPACK based blended learning model to improve engineering graduate attributes – a case study with Kirkpatrick evaluation. JARHE **17**, 421–438 (2024)
11. Waluyo, E., Nuraini: Development Instructional Design with TPACK Integrated Inquiry Learning Model to Improve Student Problem Solving Skills. Advances in Social Science, Education and Humanities Research (2021)
12. Saraç, S., Gülay Ogelman, H., Yurtseven, N., Akgün, E.: Learning to support self-regulation: empowering preschool teachers through understanding by design and blended learning. Teacher Development **29**, 510–532 (2024)
13. Yang, C.C.Y., Wu, J.-Y., Ogata, H.: Learning analytics dashboard-based self-regulated learning approach for enhancing students' e-book-based blended learning. Educ. Inf. Technol. **30**, 35–56 (2024)
14. Benson, H., Heggart, K., Williams, K.A., Smit, C., Abela, P.: Fostering Excellence in Blended Learning, a Mixed Methods Investigation of an Academic Learning Design Support Tool (2024)

Personalized Quiz Question Generation Using Knowledge Tracing and LLM

Radhika Amashi[1](✉) and Rajlaxmi Chouhan[2]

[1] SDM College of Engineering and Technology, Dharwad 580002, India
radhika9amashi@gmail.com
[2] Indian Institute of Technology, Jodhpur 342037, India

Abstract. Automatic Question Generation (AQG) for adaptive learning is an emerging field focused on creating high-quality, contextually relevant questions to support personalized assessment and learning pathways. Generating meaningful, conceptually accurate, and appropriately challenging questions tailored to diverse learners is often time-consuming and inconsistent. This study addresses this challenge by designing an adaptive AQG system that leverages Bayesian Knowledge Tracing (BKT) and Large Language Models (LLMs) to generate personalized formative assessment questions based on a student's current mastery level. The proposed system utilizes students' quiz performance data to predict concept-wise mastery through BKT, and then prompts the LLM to generate questions aligned with the predicted knowledge state. While the system successfully generates relevant questions, initial findings reveal limitations in generating diverse questions. Most questions were factual and recall-based, focusing on lower-order Bloom's Taxonomy levels, with minimal to no variation across students. There was also lack in adjusting question diversity to individual mastery levels. This study contributes to the growing literature on personalized learning by integrating knowledge tracing with LLM-driven question generation. Future enhancements will aim to address the current limitations by incorporating Bloom's Taxonomy alignment, student preferences, and prior learning history. Additionally, expanding the dataset with timestamped interactions will enable modeling of learning trajectories over time, allowing for more refined and adaptive assessment. By improving both the quality and personalization of generated questions, this approach has the potential to significantly enhance student engagement and learning outcomes in adaptive learning environments.

Keywords: Personalized Learning · Automatic Question Generation · Knowledge Tracing · LLM · Adaptive Assessments

1 Introduction

The COVID-19 pandemic has accelerated the adoption of online education and the integration of technology in higher education like engineering classrooms. Government bodies at both local and international levels have increasingly emphasized the need to incorporate educational technologies to meet the evolving expectations and needs

M. E. Auer et al. (Eds.): IMCL 2025, LNNS 1954, pp. 250–261, 2026.
https://doi.org/10.1007/978-3-032-24998-2_21

of current-generation students. For example, The National Academy of Engineering (NAE) named the development of personalized learning systems a "Grand Challenge" for the 21st century, and researchers from many different disciplines have taken aim at different features of the grand challenge they describe. EDUCAUSE (Horizon Report 2020) - projects personalized learning as a future educational development [1] and NEP 2020 highlights Use of Technology for Learning [2]. However, instructors face significant challenges in utilizing these technologies effectively particularly in generating personalized learning materials and assessments that align with each student's needs. A related concern is the rise in academic dishonesty during online assessments [3], such as quiz for formative assessments which further underscores the need for more adaptive, individualized question generation and evaluation mechanisms.

One specific and recurrent challenge for instructors is the manual creation of quiz questions for formative assessments. Developing meaningful, conceptually accurate, and appropriately challenging questions for diverse learners is time consuming and often inconsistent [4]. Automatic question generation (AQG) for adaptive learning is an evolving field that aims to create high-quality, contextually relevant questions to support personalized assessment and learning [5]. Literature shows that researchers have attempted to develop a system for automated question generation using learning analytics data, past assessments [6], from lecture notes like PPT, reading materials [7], and external knowledge bases like Wikipedia [4]. There are limited to no studies that consider students mastery level or knowledge levels for generating the personalized questions for quiz. Therefore, this study contributes to the body of literature by considering the students mastery/knowledge levels as the basis for generating personalized questions for quiz in formative assessments. This study aims to design an adaptive question generation system that leverages knowledge tracing and large language models (LLMs) with the following research question.

RQ: Can we integrate the knowledge tracing model predictions with LLM to. generate personalized objective type/quiz questions?

Knowledge tracing model a learner's understanding of concepts over time, predicting their knowledge state based on past performance and interactions. This dynamic modeling allows for a nuanced understanding of individual learning trajectories. LLMs, on the other hand, possess the ability to generate humanlike text based on vast amounts of training data. When applied to question generation, LLMs can create diverse, contextually relevant, and linguistically appropriate questions across various subjects and difficulty levels.

The integration of knowledge tracing and LLMs combines the predictive capabilities of knowledge tracing algorithms used to model and monitor a learner's knowledge state with the generative abilities of LLMs to produce quiz questions aligned with students' current understanding. By combining these technologies, an adaptive question generation system can leverage the strengths of both approaches. This approach enables educators and adaptive learning systems to tailor quiz questions to individual needs, reducing the instructor workload and efficient online assessment mechanism.

Section 2 discusses the current literature on automatic question generation systems or frameworks along with the application of knowledge tracing models in the adaptive learning systems. Section 3 discusses the methodology used to address the formulated

RQ followed by Sect. 4 discussing the results of integration of knowledge tracing and the LLM for effective question generation.

2 Literature Review

The literature review for this study is conducted in following two categories.

2.1 Automatic Question Generation:

The current state of the art in AQG leverages advanced natural language processing techniques [8], machine learning algorithms, and large language models to generate diverse and appropriate questions across various subjects and difficulty levels. [4] presents ConQuer, a concept-based quiz generation framework that uses external knowledge sources like Wikipedia and ConceptNet to improve quiz quality. This framework focuses on underlying concepts to ensure relevance and coverage of critical learning points rather than relying on keywords. The framework is evaluated using multiple dimensions of quiz quality, with LLMs serving as judges. Results show a 4.8% improvement in evaluation scores and a 77.52% win rate in pairwise comparisons over baseline quizzes. The lack of contextual relevance in such framework is addressed by exploring advanced methods for automated question generation in education, focusing on improving contextual relevance and pedagogical quality.

The author in [7] compares three approaches: In-Context Learning (ICL) using GPT-4, Retrieval Augmented Generation (RAG) using BART with a retrieval module, and a novel Hybrid Model that combines both for automated question generation. The Hybrid Model first retrieves relevant documents (RAG) and then applies few-shot learning (ICL) for question generation. Evaluations using automated and human metrics show that the Hybrid Model and ICL outperform traditional methods by producing more accurate and educationally meaningful questions. The study concludes that merging RAG and ICL enhances question generation across different cognitive levels. Question generation for specific contexts like programming, especially for students with limited prior experience in abstract reasoning and problem-solving is discussed in [6].

Early misconceptions in programming can hinder progress and lower self-confidence. To tackle this, the authors propose a framework for generating personalized prescriptive quizzes based on quiz metadata and students' past responses. These quizzes target individual misconceptions, enhancing engagement and understanding. Results show that over 91% of students felt the quizzes clarified their misconceptions and boosted their confidence. [9] paper explores targeted question generation to enhance intelligent and adaptive online education systems, which are often limited by static, hand-crafted question pools. The authors introduce LM-KT, a fine-tuned language model for deep knowledge tracing, capable of predicting a student's likelihood of correctly answering a question even for unseen questions. LM-KT is then used to guide the training of a model that generates personalized questions based on student profiles and desired difficulty levels. The approach successfully generates novel, well calibrated language translation questions for second language learners on a real educational platform.

While significant progress has been made in AQG for adaptive learning, there is a lack of studies in literature that consider student's knowledge levels or prior knowledge while question generation to cater to individual. Knowledge tracing is one concept that help in modeling or understanding the student's knowledge levels. Therefore, the next section discussed the application and utilization of knowledge tracing models in adaptive learning.

2.2 Knowledge Tracing

Knowledge tracing algorithms analyze a learner's past performance and interactions to estimate their current knowledge state and predict future performance. This dynamic modeling allows for a nuanced understanding of individual learning trajectories, enabling the identification of strengths, weaknesses, and learning patterns unique to each student. The integration of knowledge tracing in personalized education systems has shown promising results in improving learning outcomes, increasing engagement, and promoting more efficient and effective learning experiences [8].

The player knowledge modeling [10] in educational games is proposed, aiming to estimate a player's mastery of specific concepts or skills based on in game activity. The method integrates machine learning with domain-specific rules to detect when a player applies a skill and whether it was successful. These assessments are fed into a knowledge tracing framework to monitor skill mastery in real time. Ground truth data from think-aloud studies confirms the model's accuracy, showing low mean-squared error in predicting player knowledge. The results demonstrate that combining machine learning and domain knowledge can effectively trace skill application and support real-time knowledge estimation in complex educational games. The study [11] examines how Small Private Online Courses (SPOCs) in China's higher education can be enhanced using data from Learning Management Systems (LMSs). Researchers used feature engineering and nearest neighbor smoothing models to predict student performance, achieving a high ROC-AUC of 0.9390.

In Fall 2021, an engagement intervention was introduced using learning behavior dashboards, updated weekly, which led to improved student behaviors like study time, tutorial reading, and video viewing. Additionally, a personalized exercise recommendation system based on a dynamic key-value memory network was tested, showing increased concept mastery through a randomized controlled trial. Overall, the models and interventions positively impacted student learning outcomes. The paper [12] proposes a framework for evaluating short educational videos in online learning platforms using an adapted Bayesian Knowledge Tracing (BKT) model. Instead of assessing entire courses, the study focuses on how video content impacts student performance by comparing assessment outcomes with and without video interaction data. This approach provides a quantitative metric for measuring the relevance and effectiveness of videos linked to assessments. This method helped identify which videos are most helpful for struggling students, providing feedback to instructors about content needing improvement, and supports better course design.

3 Methodology

This section discusses the methodology used to address the research question (RQ), Can we integrate the knowledge tracing model predictions with LLM to generate personalized objective type/quiz questions? The novelty of this study comes from using BKT on the quiz data to generate personalized questions based on current mastery prediction of a specific student.

3.1 Data Collection

We utilized quiz data collected from a Data Science course offered to pre-final year students at a university in India. The quiz consisted of 20 questions with 4 multiple choice answers. The quiz consisted of 4 concepts, each concept having 5 questions each. Students took up this quiz as part of their assessment at the end of the semester. Quiz was designed using the google forms and was conducted in the proctored mode. A total of 125 students attempted the quiz.

Data Preprocessing: The quiz responses were exported to.csv file. The responses captured student ID, Timestamp, Questions, and the answers chosen by the student. This data was processed to make it suitable for Bayesian Knowledge Tracing (BKT) model to predict the student‘s mastery of concepts. Every question was tagged with concept name and students answers were transformed to 1 if the answer is correct and 0 if the answer is wrong. Therefore, the data consists of unique student id, concept id and the answers (in the form of 0/1).

3.2 Experimental Setup/Question Generation Pipeline

The overall question generation pipeline is shown in Fig. 1 consists of two steps. First step involves student‘s mastery predictions using state of art BKT model. Second steps involve generation of questions based on student‘s mastery produced by BKT, using LLAMA LLM.

1. Knowledge Tracing Using BKT:

   BKT estimates the student‘s latent knowledge states based on their observed performance. BKT is a specific instance of hidden markov model (hmm). It models student learning using a two-state framework, wherein a student's knowledge of a particular skill is either in a learned or unlearned state. Based on observed student responses from the quiz data, BKT employs a probabilistic update mechanism to estimate the likelihood of knowledge acquisition and predict the probability of a correct response. The BKT model predicts the probability that the student will answer the question correctly at a given step ‘correct predictions’. And, the model gives the estimated probability that the student has learned/mastered the skill at a particular point in time state predictions [13]. In this study state predictions are referred to as mastery predictions and are used for question generation. We further identify the average mastery predictions per concept for every student. A threshold of 0.6 is used to filter the low mastery concepts for every student.

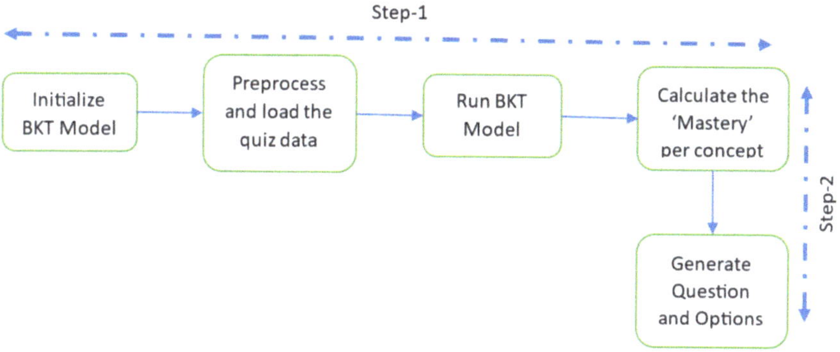

Fig. 1. Workflow of the Question Generation Pipeline

2. *Question Generation using LLAMA LLM*: The low mastery concept for every student is then passed to LLM through prompt to generate the personalized question specific to low mastery. In this study, we used open-source metallama/Llama-3-8b-chat-hf llm through together api. Following is the prompt used along with the mastery predictions.

```
prompt = (
    f"You are a helpful AI tutor. Generate a single multiple-choice question "
    f"to help a student understand the concept '{concept_id}'. "
    f"Provide exactly 4 answer choices, clearly labeled A, B, C, and D. "
    f"Indicate the correct answer at the end."
  )
```

4 Results and Discussion

This study aimed to explore whether Bayesian Knowledge Tracing (BKT) predictions can be effectively integrated with Large Language Models (LLMs) to generate personalized objective-type quiz questions for students. The novelty lies in using BKT-derived mastery estimates on concept-level quiz data to drive question generation tailored to individual students' learning needs.

4.1 Correlation Between Mastery Predictions and Student Scores

To assess the validity of BKT in predicting student's conceptual understanding, we examined the correlation between predicted mastery and actual quiz performance. A strong positive correlation of 0.91 was observed as shown in Fig. 2, indicating that the mastery levels inferred by the BKT model align well with students' real quiz scores. This supports the use of BKT as a reliable foundation for determining personalized intervention needs.

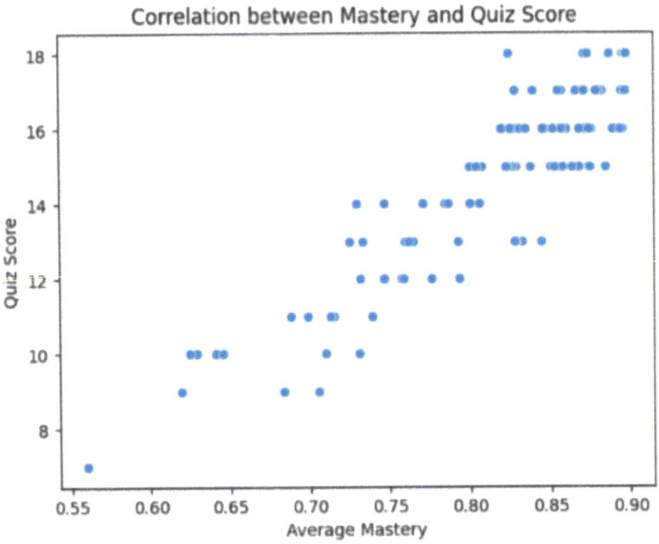

Fig. 2. Correlation between the mastery predictions and quiz scores.

4.2 BKT Predictions and Model Accuracy

The trained BKT model achieved a training accuracy of 0.804 and a Root Mean Square Error (RMSE) of 0.390, suggesting reasonable predictive performance in modeling student knowledge states across various concepts. A possible reason for the model's moderate performance could be the limited size of the dataset, as well as the absence of timestamp information within the data. Table 1 presents mastery predictions for two students (IDs: 407 and 22), with a focus on filtering concepts with mastery scores below 0.60. For instance, Student 407 shows lower mastery in Dimensionality Reduction (0.39117), Data Discretization (0.57745), and Diagnostic Analytics (0.50783). Similarly, for Student 22, Dimensionality Reduction emerges as the weakest concept. These filtered outputs serve as a direct trigger for personalized question generation.

4.3 Visualization of Student's Mastery of Concepts Through Knowledge Graphs

The predicted mastery data is further visualized through knowledge graphs as shown in Fig. 3, which maps the relationship between concepts, annotated with mastery levels. These graphs serve as an intuitive and interpretable tool for educators to understand students' conceptual gaps and learning trajectories.

Table 1. Predicted Avg Mastery of Students and Filtered Low Mastery Concepts.

Student_ID	Concept	Predicted Avg Mastery
407	Data Discretization	0.57745
	Data Science Lifecycle	0.76275
	Diagnostic Analytics	0.50783
	Dimensionality Reduction	0.39117
Filtered Concepts with Mastery < 0.60		
	Dimensionality Reduction	0.39117
	Data Discretization	0.57745
	Diagnostic Analytics	0.50783
22	Data Discretization	0.80105
	Data Science Lifecycle	0.87595
	Diagnostic Analytics	0.91476
	Dimensionality Reduction	0.39117
Filtered Concepts with Mastery < 0.60		
	Dimensionality Reduction	0.39117

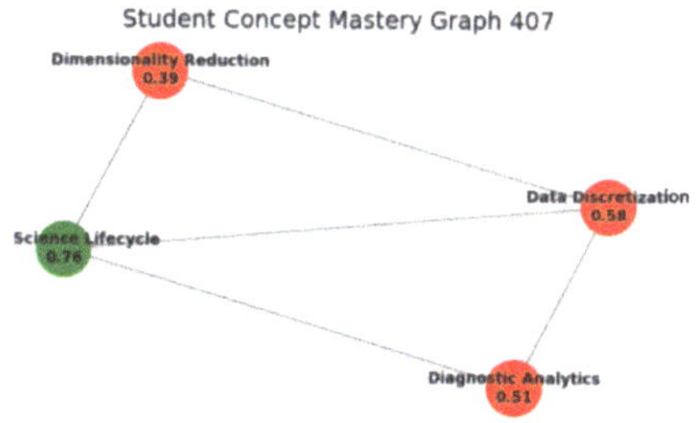

(a) Student 1: Concept Mastery

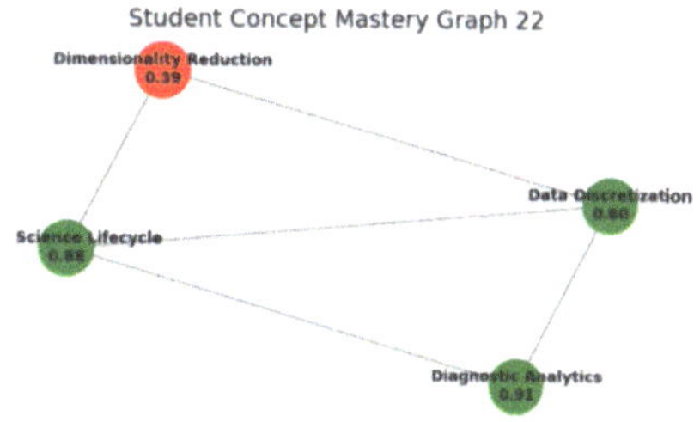

(b) Student 1: Concept Mastery

Fig. 3. Knowledge graphs and their correlation with quiz performances.

4.4 Personalized Question Generation Using LLAMA

Based on the filtered low-mastery concepts, LLAMA was prompted to generate concept-specific objective questions for each student. In the Table 2 Student 407, with low mastery in Data Discretization, received a question assessing the basic purpose of discretization in machine learning. For Dimensionality Reduction, LLAMA llm generated almost the same questions for both students (407 and 22) as shown in Table 3, with changes in the options and correct answer. This result demonstrates that there a potential of work to generate the personalized questions with appropriate difficulty and at the appropriate bloom‘s level based on students' mastery. The possible reason for this could be due to use of free llm i.e. LLAMA 3.3b chat model.

4.5 Evaluation

To evaluate the questions generated through our study, we conducted a human evaluation involving three course instructors teaching data science with 5 + year of experience. Each evaluator reviewed the nine unique questions and the multiple options with the correct answer generated. The evaluator rated each question for 4 criteria; Grammaticality, Appropriateness, Relevance and Answerability [7]. The Likert 5-point scale from 1 (worst) to 5 (best) was used for rating.

To assess how consistently the 3 raters evaluated each generated question, Fleiss's kappa was employed as the metric for interrater reliability. The obtained kappa values were 0.61 for grammaticality, 0.67 for appropriateness, 0.17 for both relevance and complexity, and 0.70 for answerability. These scores indicate a substantial degree of agreement across all the human evaluation criteria.

4.6 Discussion

While the LLM successfully generated questions aligned with the identified concepts, some of the observations from this study are, The strong positive correlation (0.91) between predicted mastery and actual quiz scores, alongside a training accuracy of 0.804 and RMSE of 0.390, indicates that BKT can reasonably model student knowledge. The questions generated through LLM based on mastery predictions lacked diversity in cognitive demand. Most were factual or recall based (lower-order Bloom's levels), with no questions generated at higher-order levels such as application or analysis. The same conceptual questions were repeated with minor wording changes across students, limiting the depth of personalization. There was no adjustment in question difficulty based on predicted mastery scores. Evaluation using Fleiss's Kappa showed substantial agreement for grammaticality (0.61), appropriateness (0.67), and answerability (0.70), but only slight agreement for relevance and complexity (0.17), suggesting a gap in generating deeper, contextually varied questions. Despite the promising results this study has few limitations, first, the dataset used was small and did not include timestamps, which limited the granularity of knowledge tracing over time. Second, the students' prior learning efforts, preferences, or historical performance were not considered in modeling or question generation and finally, contextual understanding was limited to concept-level question mapping by the course instructor; LLMs lacked richer instructional context or

Table 2. Concept-wise questions generated by llama LLM for student id: 407

Student Id: 407
Concept: Data Discretization
Q1. What is the primary goal of Data Discretization in Machine Learning? A) To reduce the dimensionality of the dataset B) To convert continuous data into a more manageable format for analysis C) To increase the accuracy of the model by reducing noise D) To handle missing values in the dataset Correct Answer: B) To convert continuous data into a more manageable format for analysis
Concept: Diagnostic Analytics
Q2. What is the primary purpose of Diagnostic Analytics in an organization? A) To identify and address potential security threats B) To analyse and improve business processes and operations C) To measure and report on employee performance and productivity D) To monitor and manage customer relationships Correct answer: B) To analyse and improve business processes and operations
Concept: Dimensionality Reduction
Q3. Question: What is the primary goal of Dimensionality Reduction in data analysis? A) To increase the number of features in the dataset B) To reduce the noise in the dataset C) To reduce the number of features in the dataset while preserving important information D) To transform the dataset into a different format Correct answer: C) To reduce the number of features in the dataset while preserving important information

Student ID:22
Concept: Dimensionality Reduction
Q1. Question: What is the primary goal of Dimensionality Reduction in data analysis? A) To add more features to the dataset B) To remove redundant or irrelevant features from the dataset C) To create a new dataset with the same number of features as the original D) To increase the accuracy of the machine learning model Correct answer: B) To remove redundant or irrelevant features from the dataset

sequencing information that could guide more refined question generation. To address these limitations and enhance the system's effectiveness, future work will focus on generating questions at multiple difficulty levels tailored to individual students' mastery levels. Incorporating student preferences, prior knowledge, and learning history into the knowledge tracing and question generation pipelines. Aligning generated questions with appropriate Bloom's Taxonomy levels, enabling the evaluation of not just factual recall but also conceptual understanding, application, and analytical thinking. Expanding the dataset and including timestamped interactions to model learning trajectories over time more effectively.

5 Conclusion

This study presents an adaptive question generation pipeline that integrates BKT with LLM to generate personalized formative assessments based on students' predicted mastery levels. By identifying low-mastery concepts through BKT and prompting an open-source LLM (LLAMA 3.3b) to generate concept specific questions, the system attempts to tailor quiz content to individual learners. Additionally, knowledge graphs were used to visualize conceptual mastery and support instructor interpretation. However, the generated questions were largely limited to lower-order Bloom's levels, with minimal diversity in cognitive demand or difficulty. Repetitive and only slightly modified questions across students indicated limited personalization. Evaluation using Fleiss's Kappa suggest a gap in generating deeper, contextually varied questions.

To improve the effectiveness of this approach, future work will focus on incorporating student preferences and, prior knowledge into both the knowledge tracing and question generation processes. Furthermore, aligning question generation with higher-order Bloom's Taxonomy levels and more cognitively diverse question sets. These enhancements aim to make adaptive question generation more pedagogically effective and genuinely personalized.

References

1. Bernacki, M.L., Greene, M.J., Lobczowski, N.G.: A systematic review of research on personalized learning: Personalized by whom, to what, how, and for wh purpose (s)?" In: Educational Psychology Review **33**(4), pp. 1675–1715. Springer (2021)
2. Kumar, A.: New education policy (NEP) 2020: A roadmap for India 2.0. In: University of South Florida (USF) M3 Publishing 3(2021), p. 36 (2021)
3. Chouhan, R.: Strategies for maintaining academic integrity in remote unproctored and proctored online assessments for engineering courses. In: Learning: Research and Practice 10(1), pp. 75–92. Taylor & Francis (2024)
4. Fu, Y., et al.: Conquer: A framework for concept-based quiz generation. In: arXiv preprint arXiv:2503.14662 (2025)
5. Nguyen, H.A., et al.: Towards generalized methods for automatic question generation in educational domains. In: European conference on technology enhanced learning, pp. 272–284. Springer (2022)
6. Soltanpoor, R., Thevathayan, C., D'Souza, D.: Adaptive remediation for novice programmers through personalized prescriptive quizzes. In: Proceedings of the 23rd annual ACM conference on innovation and technology in computer science education, pp. 51–56 (2018)

7. Maity, S., Deroy, A., Sarkar, S.: Leveraging In-Context Learning and Retrieval-Augmented Generation for Automatic Question Generation in Educational Domains. In: arXiv preprint (2025). eprint: 2501.17397
8. Al Faraby, S., Adiwijaya, A., Romadhony, A.: Review on neural question generation for education purposes. In: International Journal of Artificial Intelligence in Education 34(3), pp. 1008–1045. Springer (2024)
9. Srivastava, M., Goodman, N.: Question generation for adaptive education. In: arXiv preprint arXiv:2106.04262 (2021)
10. Kantharaju, P., et al.: Tracing player knowledge in a parallel programming educational game. In: Proceedings of the AAAI Conference on Artificial Intelligence and Interactive Digital Entertainment, Vol. 14. Issue: 1, pp. 173–179 (2018)
11. Wan, H., et al.: Pedagogical interventions in SPOCs: Learning behavior dashboards and knowledge tracing support exercise recommendation. In: IEEE Transactions on Learning Technologies 16(3), pp. 431–442. IEEE (2023)
12. MacHardy, Z.: Applications of bayesian knowledge tracing to the curation of educational videos. In: Dept. Elect. Eng. Comput. Sci., Univ. California, Berkeley, CA USA, Tech. Rep. UCB/EECS-2015-98 (2015)
13. Shen, S., et al.: A survey of knowledge tracing: Models, variants, and applications. In: IEEE Transactions on Learning Technologies. IEEE (2024)

Author Index

M. E. Auer et al. (Eds.): IMCL 2025, LNNS 1954, pp. 263–264, 2026.
https://doi.org/10.1007/978-3-032-24998-2

FSC
www.fsc.org
MIX
Papier | Fördert gute Waldnutzung
FSC® C083411

Kolibri 360 GmbH
Ferdinand-Jühlke-Straße 7
99095 Erfurt, Deutschland
produktsicherheit@kolibri360.de